JOHN DEERE

SHOP MANUAL

More information available at haynes.com
Phone: 805-498-6703

J H Haynes & Co. Ltd.
Haynes North America, Inc.

ISBN-10: 0-87288-583-6
ISBN-13: 978-0-87288-583-7

JD-62, 8S1, 14-120

Information and Instructions

This individual Shop Manual is one unit of a series on agricultural wheel type tractors. Contained in it are the necessary specifications and the brief but terse procedural data needed by a mechanic when repairing a tractor on which he has had no previous actual experience.

The material is arranged in a systematic order beginning with an index which is followed immediately by a Table of Condensed Service Specifications. These specifications include dimensions, fits, cleararances, capacities and tune-up information. Next in order of arrangement is the procedures section.

In the procedures section, the order of presentation starts with the front axle system and steering and proceeds toward the rear axle. The last portion of the procedures section is devoted to the power take-off and power lift systems.

Interspersed where needed in this section are additional tabular specifications pertaining to wear limits, torquing, etc.

How to use the index

Suppose you wnt to know the procedure for R&R (remove and reinstall) of the engine camshaft. Your first step is to look in the index under the main heading of "Engine" until you find the entry "Camshaft." Now read to the right. Under the column covering the tractor you are repairing, you will find a number which indicates the beginning paragraph pertaining to the camshaft. To locate this paragraph in the manual, turn the pages until the running index appearing on the top outside corner of each page contains the number you are seeking. In this paragraph you will find the information concerning the removal of the camshaft.

Common spark plug conditions

NORMAL

Symptoms: Brown to grayish-tan color and slight electrode wear. Correct heat range for engine and operating conditions.

Recommendation: When new spark plugs are installed, replace with plugs of the same heat range.

WORN

Symptoms: Rounded electrodes with a small amount of deposits on the firing end. Normal color. Causes hard starting in damp or cold weather and poor fuel economy.

Recommendation: Plugs have been left in the engine too long. Replace with new plugs of the same heat range. Follow the recommended maintenance schedule.

CARBON DEPOSITS

Symptoms: Dry sooty deposits indicate a rich mixture or weak ignition. Causes misfiring, hard starting and hesitation.

Recommendation: Make sure the plug has the correct heat range. Check for a clogged air filter or problem in the fuel system or engine management system. Also check for ignition system problems.

ASH DEPOSITS

Symptoms: Light brown deposits encrusted on the side or center electrodes or both. Derived from oil and/or fuel additives. Excessive amounts may mask the spark, causing misfiring and hesitation during acceleration.

Recommendation: If excessive deposits accumulate over a short time or low mileage, install new valve guide seals to prevent seepage of oil into the combustion chambers. Also try changing gasoline brands.

OIL DEPOSITS

Symptoms: Oily coating caused by poor oil control. Oil is leaking past worn valve guides or piston rings into the combustion chamber. Causes hard starting, misfiring and hesitation.

Recommendation: Correct the mechanical condition with necessary repairs and install new plugs.

GAP BRIDGING

Symptoms: Combustion deposits lodge between the electrodes. Heavy deposits accumulate and bridge the electrode gap. The plug ceases to fire, resulting in a dead cylinder.

Recommendation: Locate the faulty plug and remove the deposits from between the electrodes.

TOO HOT

Symptoms: Blistered, white insulator, eroded electrode and absence of deposits. Results in shortened plug life.

Recommendation: Check for the correct plug heat range, over-advanced ignition timing, lean fuel mixture, intake manifold vacuum leaks, sticking valves and insufficient engine cooling.

PREIGNITION

Symptoms: Melted electrodes. Insulators are white, but may be dirty due to misfiring or flying debris in the combustion chamber. Can lead to engine damage.

Recommendation: Check for the correct plug heat range, over-advanced ignition timing, lean fuel mixture, insufficient engine cooling and lack of lubrication.

HIGH SPEED GLAZING

Symptoms: Insulator has yellowish, glazed appearance. Indicates that combustion chamber temperatures have risen suddenly during hard acceleration. Normal deposits melt to form a conductive coating. Causes misfiring at high speeds.

Recommendation: Install new plugs. Consider using a colder plug if driving habits warrant.

DETONATION

Symptoms: Insulators may be cracked or chipped. Improper gap setting techniques can also result in a fractured insulator tip. Can lead to piston damage.

Recommendation: Make sure the fuel anti-knock values meet engine requirements. Use care when setting the gaps on new plugs. Avoid lugging the engine.

MECHANICAL DAMAGE

Symptoms: May be caused by a foreign object in the combustion chamber or the piston striking an incorrect reach (too long) plug. Causes a dead cylinder and could result in piston damage.

Recommendation: Repair the mechanical damage. Remove the foreign object from the engine and/or install the correct reach plug.

SHOP MANUAL
JOHN DEERE

MODELS 670, 770, 870, 970 and 1070

The 13 digit product identification (serial) number is on a plate (Fig. 1) located above the rear pto output shaft on Models 670 and 770 or below the rear pto shaft on Models 870, 970 and 1070. The engine serial number plate (Fig. 2) is attached to the top of the rocker arm cover.

Fig. 1—The tractor serial number plate is located just below rear pto output shaft. Illustration on left is typical of 670 and 770 models; illustration on right is typical of 870, 970 and 1070 models.

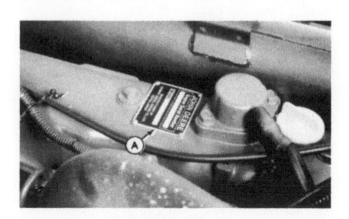

Fig. 2—Engine model and serial numbers are located on plate attached to top of rocker arm cover. All models are similar.

INDEX (By Starting Paragraph)

**Models
670, 770, 870,
970 and 1070**

**Models
670, 770, 870,
970 and 1070**

DUAL DIMENSIONS

This shop manual provides specifications in both U.S. Customary and Metric (SI) systems of measurement. The first specification is given in the measuring system perceived by us to be the preferred system when servicing a particular component, while the second specification (given in parenthesis) is the converted measurement. For instance, a specification of 0.011 inch (0.28 mm) would indicate that we feel the preferred measurement in this instance is the U.S. Customary system of measurement and the Metric equivalent of 0.011 inch is 0.28 mm.

CONDENSED SERVICE DATA

			Models		
	670	**770**	**870**	**970**	**1070**
GENERAL					
Engine Make			Yanmar		
Engine Model	3TNA72-UJX	3TN82-RJX	3TN84-RJX	4TN842-RJX	4TN84-RJX
Number of Cylinder		3		4	
Bore	72 mm (2.83 in.)	82 mm (3.23 in.)	84 mm (3.31 in.)	82 mm (3.23 in.)	84 mm (3.31 in.)
Stroke	72 mm (2.83 in.)		86 mm (3.39 in.)		
Displacement	879 mm (53.6 cu. in.)	1362 mm (83.1 cu. in.)	1431 mm (87.3 cu. in.)	1816 mm (110.8 cu. in.)	1906 mm (116.3 cu. in.)
Compression Ratio	22.3:1	18.1:1	17.8:1	18.1:1	17.8:1
Fuel Injection Type	Indirect		Direct		
Injection Pump Type			Inline		
Battery —					
Voltage			12		
Terminal Grounded			Negative		
Transmission —					
Standard	Sliding gear	Sliding gear	Collar shift	Collar shift	Synchronized
Optional	...	...	Synchronized	Synchronized	...
Speeds	8 Forward & 2 Reverse		9 Forward & 2 Reverse *		

* High speed range is blocked out of some models when main transmission is in reverse.

TUNE-UP					
Firing Order **		1-3-2		1-3-4-2	
Valve Clearance, Cold —					
Inlet			0.2 mm (0.008 in.)		
Exhaust			0.2 mm (0.008 in.)		
Valve Face and Seat Angle —					
Inlet			30°		
Exhaust			45°		
Compression Pressure —					
Minimum			2448 kPA (355 psi)		
Injector —					
Opening Pressure	11,242-12,202 kPa (1630-1770 psi)		19,120-20,080 kPa (2773-2913 psi)		
Engine Low Idle			925 rpm		
Engine High Idle	3025 rpm		2775 rpm		2875 rpm
Engine Rated Speed	2850 rpm		2600 rpm		
Engine Power					
Rating @ Pto	11.9 kW (16 hp)	14.9 kW (20 hp)	18.6 kW (25 hp)	22.4 kW (30 hp)	26.1 kW (35 hp)

** Cylinders are numbered from back to front. Number 1 cylinder is rear cylinder.

SIZES					
Crankshaft Main					
Journal Diameter	43.97-43.98 mm (1.731-1.732 in.)		49.952-49.962 mm (1.9666-1.9670 in.)		
Crankshaft Crankpin —					
Diameter	39.97-39.98 mm (1.5736-1.574 in.)		47.952-47.962 mm (1.8879-1.8883 in.)		

CONDENSED SERVICE DATA (CONT.)

	Models				
	670	**770**	**870**	**970**	**1070**
Piston Pin Diameter	20.991-21.0 mm (0.826-0.827 in.)		25.987-26.0 mm 1.023-1.024 in.)		
Valve Stem Diameter — Inlet	6.945-6.960 mm (0.273-0.274 in.)		7.960-7.975 mm (0.313-0.314 in.)		
Exhaust	6.945-6.960 mm (0.273-0.274 in.)		7.960-7.975 mm (0.313-0.314 in.)		

CLEARANCES

Main Bearing, Diametral Clearance	0.020-0.0072 mm (0.0008-0.0028 in.)		0.038-0.093 mm (0.0015-0.0037 in.)		
Rod Bearing Diametral Clearance	0.020-0.0072 mm (0.0008-0.0028 in.)		0.038-0.090 mm (0.0015-0.0035 in.)		
Camshaft Bearing, Diametral Clearance — Front	0.040-0.085 mm (0.0016-0.0033 in.)		0.040-0.130 mm (0.0016-0.0051 in.)		
Crankshaft End Play			0.09-0.271 mm (0.004-0.011 in.)		

CAPACITIES

Cooling System	.8 L (4 qt.)	4.8 L (5.1 qt.)	5.2 L (5.5 qt.)	5.8 L (6.1 qt.)	5.8 L (6.1 qt.)
Crankcase With Filter	2.6 L (2.7 qt.)	4.0 L (4.2 qt.)	4.8 L (5.1 qt.)	5.3 L (5.6 qt.)	5.3 L (5.3 qt.)
Transmission	15 L (4.0 gal.)		21 L (5.5 gal.)		
Front Drive Axle			2.13 L (2.25 qt.)		

FRONT AXLE SYSTEM
(TWO-WHEEL DRIVE)

FRONT AXLE ASSEMBLY AND
STEERING LINKAGE

Two-Wheel Drive Models

1. WHEELS AND BEARINGS. Front wheel bearings should be removed, cleaned, inspected and renewed or repacked every 500 hours of operation. To remove front wheel hub and bearings, raise and support the front axle, then unbolt and remove the tire and wheel assembly. Remove cap (1—Fig. 3 or Fig. 4), cotter pin, castellated nut (2), washer (3) and outer bearing cone (4). Slide the hub assembly from spindle axle shaft. Remove seal (8) and cone of inner bearing (7).

Inspect bearings for scratches, score marks, pitting or other damage. Drive bearing cups from hub (6) if renewal is required. Pack wheel bearings liberally with a suitable wheel bearing grease.

Reassemble by reversing disassembly procedure. Tighten castellated nut (2) until a slight drag is felt, back nut off ¼ turn or just enough to install cotter pin, then install cap (1). Tighten front wheel retaining

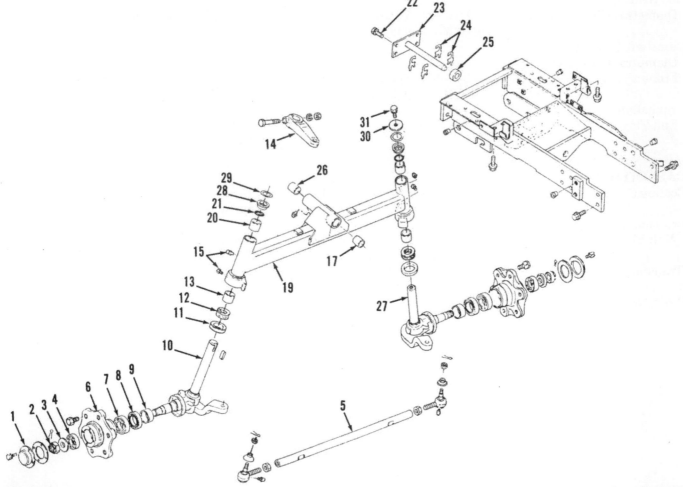

Fig. 3—Exploded view of non-adjustable front axle typical of some two-wheel drive models. Refer to Fig. 4 for adjustable axle.

1. Cap	9. Seal ring	19. Axle	25. Spacer
2. Castellated nut	10. Left spindle	20. Upper bushing	26. Bushing (same as 17)
3. Washer	11. Seal	(same as 13)	27. Right spindle
4. Outer bearing	12. Thrust washer	21. Seal ring	28. Retainer
5. Tie rod	13. Bearing (same as 20)	22. Screws	29. Shims
6. Hub	14. Steering arm	23. Pivot pin	30. Washer
7. Inner bearing	15. Grease fittings	24. Shims	31. Screw
8. Seal	17. Bushing (same as 26)		

screws of 670 and 770 models to 133 N•m (98 ft.-lbs.) torque. Tighten front wheel retaining screws to 217 N•m (160 ft.-lbs.) torque for all 870, 970 and 1070 models. Check wheel retaining screws for correct torque frequently.

2. TIE ROD AND TOE-IN. A single tie rod (5—Fig. 3 or Fig. 4) connects left and right steering arms of

spindles (10 and 27). Automotive type tie rod ends are not adjustable for wear and should be renewed if worn. Rod ends are threaded into tie rod. Front wheel toe-in is changed by turning rod ends into tie rod to adjust the distance between ends. Recommended toe-in is 3-9 mm (⅛-⅜ in.) and should be measured between wheel rims on centerline of axle, parallel to ground. Rotate wheels and re-measure to be sure that

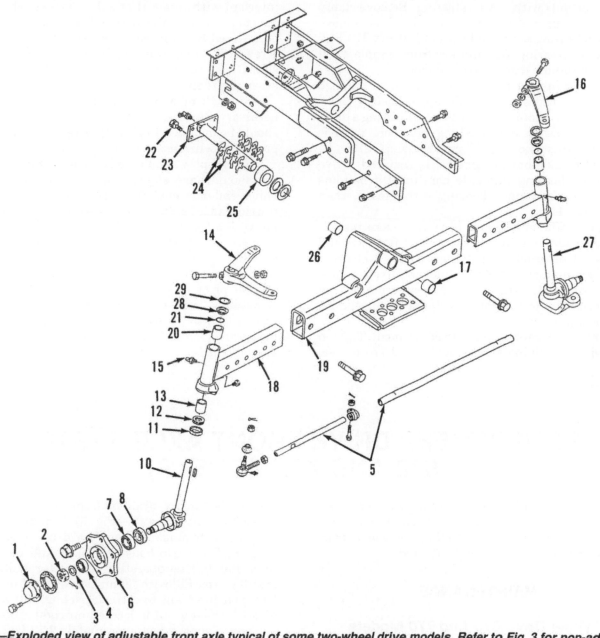

Fig. 4—Exploded view of adjustable front axle typical of some two-wheel drive models. Refer to Fig. 3 for non-adjustable axle.

1. Cap	8. Seal	16. Steering arm	23. Pivot pin
2. Castellated nut	10. Left spindle	17. Bushing (same as 26)	24. Shims
3. Washer	11. Seal	18. Axle extension	25. Spacer
4. Outer bearing	12. Thrust washer	19. Axle center member	26. Bushing (same as 17)
5. Tie rod & tube	13. Bearing	20. Upper bushing	27. Right spindle
6. Hub	14. Steering arm	21. Seal ring	28. Retainer
7. Inner bearing	15. Grease fittings	22. Screws	29. Shims

wheels are not bent, giving incorrect reading. Tighten rod end jam nut after toe-in is correctly set.

On models with adjustable tread width axle (Fig. 4), clamp is used for quick adjustment of tie rod length when changing tread width. Toe-in is set by turning tie rod ends the same as nonadjustable axles.

3. SPINDLES AND BUSHINGS. To remove spindle (10 or 27—Fig. 3 or Fig. 4), first remove wheel and hub, then detach tie rod ends from steering arms. Steering cylinder must be disconnected from steering arms of models with power steering. Remove clamp screw from upper steering arm (14 or 16) or remove spindle retaining screw and washer (30 and 31—Fig. 3). Remove steering arm and key from spindle, then withdraw spindle from axle of all models.

Clean and inspect parts for wear or other damage and renew as necessary. Spindle upper bushing (20—Fig. 3 or Fig. 4) should be pulled from top using a blind hole puller. Drive new bushing into bore until it bottoms against shoulder. Thrust bearing (12) can be pulled from axle bore if renewal is required. Drive new thrust bearing into axle bore until it bottoms against shoulder, pack bearing with grease, then install seal (11).

Clean spindle and grease bearings before installing, then insert spindle into axle bore. Install key and upper steering arms, then attach rod ends to steering arms. Tighten castellated nuts and install cotter pins. Attach power steering cylinder to steering arm of models so equipped, tighten castellated nut and install cotter pin.

Refer to paragraph 2 for toe-in adjustment. Tighten front wheel retaining screws of 670 and 770 models to 133 N·m (98 ft.-lbs.) torque. Tighten front wheel retaining screws to 217 N·m (160 ft.-lbs.) torque for all 870, 970 and 1070 models. Check wheel retaining screws for correct torque frequently.

4. AXLE MAIN MEMBER, PIVOT PIN AND BUSHINGS. To remove front axle assembly, first remove any front mounted equipment and weights. On models with power steering, detach both ends of steering cylinder, but leave hoses attached. Raise and block front of tractor in such a way that it will not interfere with removal of axle. Remove wheels and support axle in center with a floor jack so that it can be lowered and moved away from tractor. Remove screws (22—Fig. 3 or Fig. 4) and carefully remove shims (24). Remove pivot pin (23), lower the axle assembly, then roll axle from under tractor.

Check axle pivot bushings (17 and 26) and renew if necessary. Bushings are pressed into bore of axle and should be installed flush with bore. Reverse removal procedure when assembling.

Axle end play should be less than 1 mm (0.04 in.), but should have slight clearance. To measure and adjust end play, push the axle to the rear, then measure clearance between front of axle and plate on pivot pin with a feeler gauge. Shims (24) should be added to increase clearance. Make sure that screws (22) attaching plate (23) are tight when measuring, but that axle does not bind. Washers are located at ends of axle pivot of some models. If washers are not reinstalled, end play may be excessive. Tighten screws (22) to 88 N·m (65 ft.-lbs.) torque. Refer to paragraphs 2 and 3 for additional torque values and assembly notes.

FOUR-WHEEL DRIVE FRONT AXLE SYSTEM (670 AND 770 MODELS)

5. Mechanical front-wheel drive is available for all models. There may be differences between the front-wheel drive systems as noted in the servicing instructions which follow for 670 and 770 models.

MAINTENANCE

Four-Wheel Drive 670 And 770 Models

6. Oil should be maintained at top mark of dipstick attached to fill plug located on right side of axle housing. To check oil level remove dipstick and wipe clean. Insert dipstick, but do not screw dipstick into axle housing. Remove dipstick and check oil level on dipstick.

Every 500 hours, drain oil from front drive axle by removing vent plugs (V—Fig. 5) from both sides of axle and drain plugs (D) from final drive housings and plug (P) from axle housing. Oil should be drained immediately after operating the tractor while the oil is still warm. Fill with "John Deere GL-5" or equivalent gear lubricant to correct level. Oil fill plug and dipstick (F—Fig. 6) is located on right side of axle housing. Capacity is approximately 2.13 L (2.25 qt.); however, some lubricant may be trapped in housing.

Check screws retaining front wheels occasionally to make sure that they do not loosen. Tighten front wheel retaining screws of 670 and 770 models to 133 N·m (98 ft.-lbs.) torque.

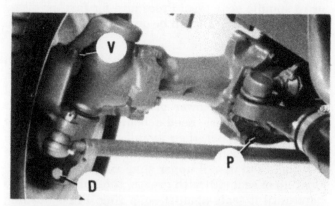

Fig. 5—Oil can be drained from mechanical front-wheel drive axle used on 670 and 770 models by removing both drain plugs (D) and drain plug (P). Removing vent plug (V) located in the top of final drive housing will allow the oil to drain quicker. Dipstick is attached to fill plug (F—Fig. 6) for measuring oil level.

FRONT AXLE ASSEMBLY AND STEERING LINKAGE

Four-Wheel Drive 670 And 770 Models

7. TIE ROD AND TOE-IN. A single tie rod connects left and right steering arms. Automotive type tie rod ends are not adjustable for wear and should be renewed if worn. Attach steering tie rod ends to steering arms, tighten castellated nuts to 68 N·m (50 ft.-lbs.) torque and install cotter pins.

Recommended toe-in is 3-9 mm (1/8-3/8 in.) and is adjusted by threading rod ends into tie rod. To check alignment, measure between wheel rims on center-line of axle, parallel to ground. Rotate wheels and re-measure to be sure that wheels are not bent, giving incorrect reading. Tighten rod end jam nut after wheel alignment is correctly set.

8. R&R AXLE ASSEMBLY. To remove front axle assembly, first remove any front mounted equipment and weights. Disconnect battery ground and detach both ends of steering cylinder, but leave hoses attached. Raise and block front of tractor in such a way that it will not interfere with removal of axle. Support tractor behind axle and remove both front wheels. If unit is to be disassembled, remove plugs and drain oil from front axle assembly. Remove the front axle driveshaft. Support axle level with floor to prevent tipping and in such a way that it can be lowered and moved away from tractor. Remove screws retaining the axle pivot pin (51—Fig. 6) and carefully remove shims (50). Remove pivot pin (51), lower the axle assembly, then roll axle safely from under tractor.

Check axle pivot bushings (26—Fig. 6) and renew if necessary. Bushings are pressed into bore of housing (7) and should be installed flush with bore. Reverse removal procedure when assembling. Axle end

play should be 0.127-1.016 mm (0.005-0.040 in.). To measure and adjust end play, push axle to the rear, then measure clearance between front of axle and plate on pivot pin with a feeler gauge. Shims (50) should be added to increase clearance. Make sure that screws attaching pivot pin plate (51) are tight when measuring, but that axle does not bind. Tighten screws to 88 N·m (65 ft.-lbs.) torque. Attach steering cylinder to right steering arm, tighten castellated nuts to 128 N·m (94 ft.-lbs.) torque and install cotter pins. Tighten front wheel retaining screws of 670 and 770 models to 133 N·m (98 ft.-lbs.) torque. Tighten front wheel retaining screws to 217 N·m (160 ft.-lbs.) torque for 870 model. Tighten front wheel retaining screws of 970 and 1070 models to 190 N·m (140 ft.-lbs.) torque. Refer to paragraph 7 for additional torque values and assembly notes.

FRONT FINAL DRIVE

Four-Wheel Drive 670 And 770 Models

9. R&R AND OVERHAUL. Raise and block front of tractor in a way that is safe and will not interfere with the removal of front wheels or final drive. Remove vent plug (V—Fig. 5) and drain plug (D) and drain oil from axle housing. Remove the six screws that attach final drive housing (12—Fig. 6) to spindle housing (2), then separate housings. Bearing (21) will probably remain in spindle housing, but can be removed with puller. Outer race of bearing (21) is located in bore of gear (19). Wheel axle and hub (11) can be pressed from gear (19) after removing snap ring (20). Wear sleeve (14) can be removed if renewal is required. Gear (19) can be pressed from bearing (17), then bearing (17) can be pressed from housing after removing seal (15) and snap ring (16). Be sure to remove old gasket (13) and install new gasket.

Press new bearing (17) against snap ring (18), then install snap ring (16). Coat lip of seal (15) and press into housing bore. Press wheel axle and hub (11) into gear (19), then install snap ring (20). Install drain plug (D) and new gasket (13), then insert wheel axle and housing into spindle housing (2). Tighten retaining screws in a crossing pattern to 51 N·m (39 ft.-lbs.) torque. Fill axle with "John Deere GL-5" or equivalent gear lubricant to correct level marked on dipstick of fill plug (F—Fig. 6). Capacity is approximately 2.13 L (2.25 qt.); however, some lubricant may be trapped in housing. Tighten front wheel retaining screws to 133 N·m (98 ft.-lbs.) torque.

FRONT DRIVE SPINDLE

Four-Wheel Drive 670 And 770 Models

10. R&R AND OVERHAUL. To remove the front drive spindle assembly, first refer to paragraph 9 and remove front final drive. Disconnect tie rod end from steering arm and, if right spindle is being removed, detach steering cylinder. Remove six screws which attach spindle housing (3—Fig. 6) to axle housing (7) and remove the complete spindle assembly. Driveshafts (48) can be withdrawn if required.

Remove the two nuts and two screws which attach steering arm (4) to spindle gear case (2). Use a soft hammer to bump steering arm from dowel pins. Remove wear sleeve (34) from shaft of steering arm only if new wear sleeve is to be installed. Seal (35) can be pulled from bore in spindle housing.

Bump spindle gear case (2) from spindle housing (3). Spindle shaft (30) may remain in spindle housing

or in spindle gear case. Procedure for further disassembly will vary, but will be obvious. Refer to Fig. 6.

When assembling, press bearing (27) into bottom bore of gear case (2), then position washer (28) and gear (29) in case above the installed bearing. Install snap ring (41) in lower groove, followed by lower washer (42). Press needle bearing (43) into bore against the washer, then install inner race (44). Install upper washer (42) above the installed needle bearing, then install upper snap ring (41) in groove. Coat lip of seal (40) with grease, then press seal into bore until it seats against snap ring. Press bushing (39) into bore against seal.

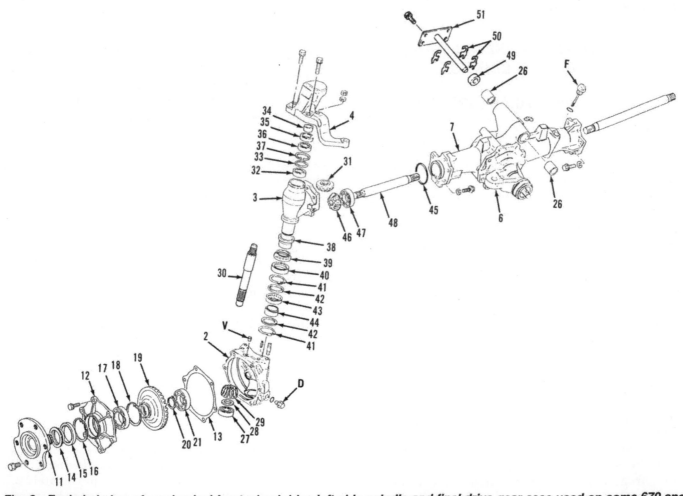

Fig. 6—Exploded view of mechanical front-wheel drive left side spindle and final drive gear case used on some 670 and 770 models. Parts for right side are similar.

2. Spindle gear case	18. Snap ring	32. Ball bearing (same as 36)	42. Washers (2 used)
3. Spindle housing	19. Bevel gear	33. Washer	43. Needle bearing
4. Left steering arm	20. Snap ring	34. Oil seal wear sleeve	44. Inner race
11. Wheel hub & axle	21. Ball bearing	35. Seal	45. Seal ring
12. Final drive housing	26. Bushings	36. Ball bearing (same as 32)	46. Drive gear
13. Gasket	27. Ball bearing	37. Snap ring	47. Ball bearing
14. Sleeve	28. Washer	38. Oil seal wear sleeve	48. Left drive shaft
15. Seal	29. Final drive pinion	39. Bushing	49. Spacer
16. Snap ring	30. Spindle shaft	40. Oil seal	50. Shims
17. Ball bearing	31. Driven gear	41. Snap rings (2 used)	51. Pivot pin

Carefully press oil seal wear sleeve (38) onto spindle housing (3) until fully seated. Press bearing (32) into bore until it is just below snap ring groove, then install washer (33) and snap ring (37). Press bearing (36) into bore against snap ring, grease lip of seal (35), then press seal into bore against bearing. Oil seal wear sleeve (34) is pressed onto post of steering arm (4).

Position upper gear (31) in spindle housing (3), then insert spindle shaft (30) through spindle housing with shorter splines in gear (31). Upper end of shaft should enter bearing (32). Tap shaft to seat shaft in bearing and gear.

Make sure that lower gear (29) and washer (28) are correctly positioned over the installed bearing (27), then insert spindle shaft and housing assembly into gear housing (2). Make sure that shaft splines are aligned with gear and end of shaft enters the installed lower bearing (27). Install steering arm (4), but do not tighten retaining nuts and screws yet. Insert drive gear (46) and bearing (47) into bore of spindle housing (3) and install new seal ring (45).

If removed, install driveshaft (48) with short splines out toward gear (46). Position spindle assembly and bearing (47) against axle housing, then install retaining screws. Tighten the six screws retaining spindle to axle to 51 N·m (39 ft.-lbs.) torque, then tighten the two nuts and two screws retaining steering arm to 51 N·m (39 ft.-lbs.) torque.

Install bearing (21), drain plug (D) and new gasket (13). Insert wheel axle and housing into spindle housing (2). Tighten retaining screws in a crossing pattern to 51 N·m (39 ft.-lbs.) torque. Fill axle with "John Deere GL-5" or equivalent gear lubricant to correct level marked on dipstick of fill plug (F—Fig. 6). Capacity is approximately 2.13 L (2.25 qt.); however,

some lubricant may be trapped in housing. Tighten front wheel retaining screws to 133 N·m (98 ft.-lbs.) torque.

DIFFERENTIAL AND INPUT BEVEL GEARS

Four-Wheel Drive 670 And 770 Models

11. R&R AND OVERHAUL. The differential assembly can be removed without removing the front drive axle housing (7—Fig. 6) from tractor, but some mechanics prefer to remove the complete axle assembly for ease of service and for safety. Refer to paragraph 8 for removal of front drive axle.

To remove the differential assembly (Fig. 7), first raise and block front of tractor in a way that is safe and will not interfere with removal of differential. Remove the driveshaft. Remove drain plugs (D—Fig. 5) and drain oil from axle housing, then unbolt and remove spindle housings (3—Fig. 6) from both ends of axle. Withdraw shafts (48) from differential, then unbolt and remove differential housing (49—Fig. 7) from axle housing (7). It may be necessary to bump differential housing (49) from locating dowel pins.

The differential and ring gear can be removed as an assembly from housing (49) after removing snap rings (64). **Be careful not to lose or damage shims (75) which set backlash between bevel pinion (62) and ring gear (71).** The input shaft and bevel gear (62) can be removed after removing snap rings (54 and 60), then pressing shaft and bearing (61) from housing and bearing (57).

Use puller to remove bearings (65 and 74). Ring gear (71) is attached to differential case (73) with six

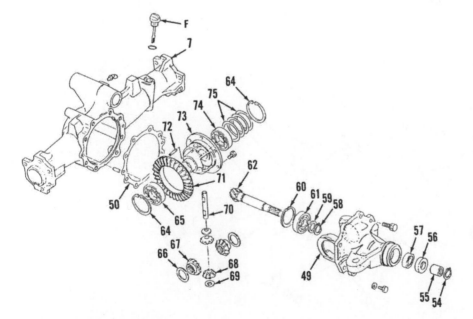

Fig. 7—Exploded view of front axle differential and input bevel gears, typical of 670 and 770 models so equipped.

49. Differential carrier housing	65. Ball bearing (same as 74)
50. Gasket	66. Thrust washer
54. Snap ring	67. Side gear
55. Wear sleeve	68. Pinion (2 used)
56. Oil seal	69. Thrust washer (2 used)
57. Ball bearing	70. Cross shaft
58. Snap ring	71. Ring gear
59. Spacer	72. Roll pin
60. Snap ring	73. Differential housing
61. Ball bearing	74. Ball bearing (same as 65)
62. Drive pinion & shaft	75. Shims
64. Snap rings	

screws. Drive roll pin (72) from case and cross shaft (70), then drive cross shaft from case and gears (68). Roll gears (68) and thrust washers (69) from differential case, then withdraw side gears (67) and thrust washers (66).

Assembly of differential is reverse of disassembly procedure. Apply medium strength thread lock to screws retaining ring gear (71) and tighten to 80 N•m (59 ft.-lbs.) torque.

If removed, press new bearing (61) onto bevel pinion shaft (62), install spacer (59) and snap ring (58). Press bearing (57) into housing (49), then press input shaft (62) and bearing (61) into housing and bearing (57). Install snap ring (60), seal (56), sleeve (55) and snap ring (54). After bevel pinion is installed, install differential assembly with ring gear on right side of pinion as shown.

Make sure that bearings (65 and 74) are firmly seated against differential case (73), then install differential in housing (49). Install shims (75) that were removed when installing differential, install snap rings (64), then measure backlash between ring gear (71) and pinion (62). Install sufficient thickness of shims (75) to limit backlash to 0.08-0.13 mm (0.003-0.005 inch) when measured at outer edge of ring gear teeth.

Coat gasket (50) with sealer and install differential, drive pinion and housing assembly, tightening retaining screws evenly to 108 N•m (80 ft.-lbs.) torque. Install both left and right driveshafts (48), then install final drives and steering spindles as outlined in paragraph 10. If removed, refer to paragraph 8 and install the axle assembly.

Install drain plugs and fill axle with "John Deere GL-5" or equivalent gear lubricant to correct level marked on dipstick of fill plug (F—Fig. 7). Capacity is approximately 2.13 L (2.25 qt.); however, some lubricant may be trapped in housing. Tighten front wheel retaining screws to 133 N•m (98 ft.-lbs.) torque.

FRONT-WHEEL DRIVE GEAR CASE

Four-Wheel Drive 670 And 770 Models

12. R&R AND OVERHAUL. Power for front wheel drive is from a gear located on front of main drive bevel pinion shaft, which drives gear (15—Fig. 8). To remove front-wheel-drive shaft (12) and related parts, it is necessary to split clutch housing from the transmission housing as outlined in paragraph 143. Remove lock plate (7) and push shift arm (5) away from engagement coupling (14). Withdraw output

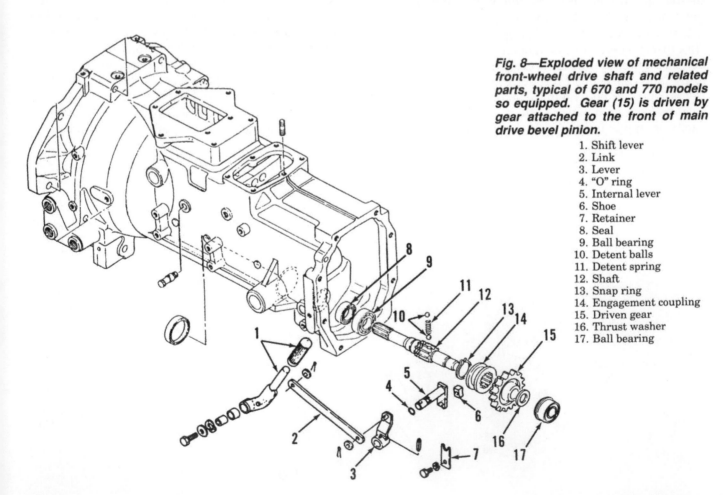

Fig. 8—Exploded view of mechanical front-wheel drive shaft and related parts, typical of 670 and 770 models so equipped. Gear (15) is driven by gear attached to the front of main drive bevel pinion.

1. Shift lever
2. Link
3. Lever
4. "O" ring
5. Internal lever
6. Shoe
7. Retainer
8. Seal
9. Ball bearing
10. Detent balls
11. Detent spring
12. Shaft
13. Snap ring
14. Engagement coupling
15. Driven gear
16. Thrust washer
17. Ball bearing

shaft assembly. Note that detent balls (10) will fly out of output shaft when engagement coupling and driven gear are removed. Front seal (8) must be removed and installed from inside case, with spring-loaded lip toward inside.

When assembling, reverse disassembly procedure. Reconnect transmission to clutch housing as outlined in paragraph 143. Fill transmission, rear axle center housing and hydraulic system as outlined in paragraph 178.

FOUR-WHEEL DRIVE FRONT AXLE SYSTEM (870 MODEL)

13. Mechanical front-wheel drive is available. Refer to the servicing instructions which follow for 870 model.

MAINTENANCE

Four-Wheel Drive 870 Model

14. Oil should be maintained at top mark of dipstick attached to fill plug (F—Fig. 9). Every 500 hours, drain oil from front drive axle by removing the two plugs (D) and plug (P) from housings. Fill with "John Deere GL-5" or equivalent gear lubricant to correct level. Capacity is approximately 2.13 L (2.25 qt.); however, some lubricant may be trapped in housing.

Check screws retaining front wheels occasionally to make sure that they do not loosen. Tighten front wheel retaining screws to 217 N·m (160 ft.-lbs.) torque.

FRONT AXLE ASSEMBLY AND STEERING LINKAGE

Four-Wheel Drive 870 Models

15. TIE ROD AND TOE-IN. A single tie rod connects left and right steering arms. Automotive type ends are not adjustable for wear and should be renewed if worn. Attach steering tie rod ends to steering arms, tighten castellated nuts to 68 N·m (50 ft.-lbs.) torque and install cotter pins.

Recommended toe-in is 3-9 mm (⅛-⅜ in.) and is adjusted by threading rod ends into tie rod to change the distance between ends. To check alignment, measure between wheel rims on centerline of axle, parallel to ground. Rotate wheels and re-measure to be sure that wheels are not bent, giving incorrect reading. Tighten rod end jam nut after wheel alignment is correctly set.

16. R&R AXLE ASSEMBLY. To remove front axle assembly, first remove any front mounted equipment and weights. Detach both ends of steering cylinder. Hoses to steering cylinder can remain attached if cylinder is moved and safely supported while detached from front axle. Raise and block front of tractor in such a way that it will not interfere with removal of axle. Support tractor behind axle and remove both front wheels. If axle unit is to be disassembled, remove plugs (D and P—Fig. 10) to drain oil from front axle assembly. Remove the front axle driveshaft. Support axle level with floor to prevent tipping and in such a way that it can be lowered and moved away from tractor. Remove screws retaining the axle pivot pin (81—Fig. 11), then carefully remove shims (80). Remove pivot pin (81), lower the axle assembly and roll axle safely from under tractor.

Check axle pivot bushings (26—Fig. 11) and renew if necessary. Install new bushings into bore of housing (7) using suitable bushing driver. Reverse removal procedure when assembling. Note that pivot pin

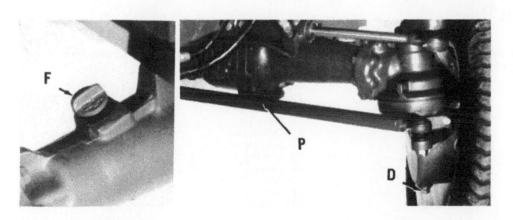

Fig. 9—View of drain plugs (D & P). Filler plug and attached dipstick (F) is on right side of axle.

grease fitting, if used, should point downward when pin is installed. Axle end play should be 0.127-1.016 mm (0.005-0.040 in.). To measure and adjust end play, push axle to the rear, then measure clearance between front of axle and plate on pivot pin with a feeler gauge. Shims (80—Fig. 11) should be added to increase clearance. Make sure that screws attaching pivot pin plate (81) are tight when measuring, but that axle does not bind. Screws should be tightened to 88 N·m (65 ft.-lbs.) torque. Attach tie rod and steering cylinder or drag link to steering arm, tighten castellated nuts to 68 N·m (50 ft.-lbs.) torque and install cotter pins. Tighten front wheel retaining

screws to 217 N·m (160 ft.-lbs.) torque. Refer to paragraph 14 and 15 for additional assembly notes.

FRONT FINAL DRIVE AND SPINDLE

Four-Wheel Drive 870 Models

17. REMOVE AND REINSTALL. Raise and block front of tractor in a way that is safe and will not interfere with the removal of front wheels or final drive. Remove plugs (D and P—Fig. 10) and drain oil from axle housing. Detach tie rod, drag link and

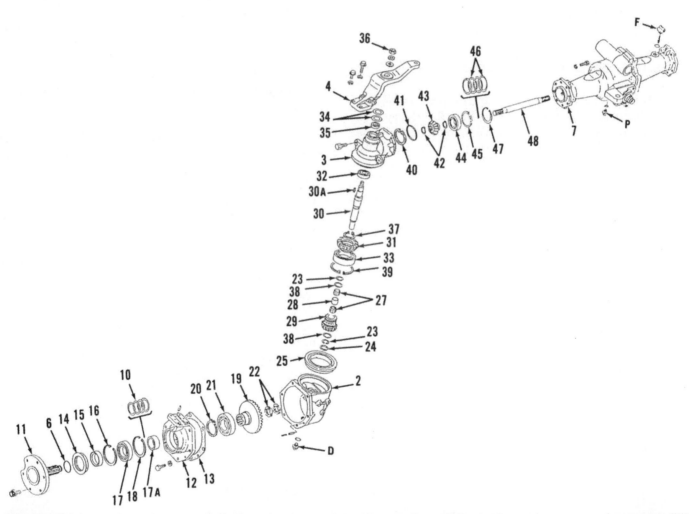

Fig. 10—Exploded view of mechanical front-wheel drive left side spindle and final drive gear case used on some 870 models. Parts for right side are similar.

2. Spindle gear case	17. Ball bearing	28. Spacer	38. Washers
3. Spindle housing	17A. Spacer	29. Final drive pinion	39. Snap ring
4. Left steering arm	18. Snap ring	30. Spindle shaft	40. Snap ring
6. "O" ring	19. Bevel gear	30A. Drive key	41. "O" ring
10. Shims	20. Snap ring	31. Driven gear	42. Snap rings
11. Wheel hub & axle	21. Ball bearing	32. Bearing	43. Drive gear
12. Final drive housing	22. Nuts	33. Bearing	44. Bearing
13. Gasket	23. Bushing	34. Shims	45. Spacer
14. Seal	24. Spacer	35. Seal	46. Shims
15. Sleeve	25. Seal	36. Nut	47. Snap ring
16. Snap ring	27. Bearings	37. Snap ring	48. Left drive shaft

steering cylinder from steering arm, depending upon options and which final drive is being removed. Support final drive assembly safely with a floor jack so that unit can be withdrawn, then remove the eight screws which attach housing (3) to housing (7). Carefully separate housings, then carefully move final drive away from tractor.

Remove old "O" ring (41), then lubricate new "O" ring with petroleum jelly and position on axle housing. Install the final drive and spindle, tightening retaining screws to 51 N·m (39 ft.-lbs.) torque. "Loctite" or similar thread lock should be applied to threads of retaining screws. Connect tie rod, drag link and steering cylinder to steering arm and tighten nuts to 68 N·m (50 ft.-lbs.) torque. Tighten front

wheel retaining screws to 217 N·m (160 ft.-lbs.) torque. Refer to paragraph 14 and fill axle to correct level with lubricant.

18. OVERHAUL. Remove snap ring (47—Fig. 10 or Fig. 11), then pull shaft (48 or 63), gear (43), spacer, shims and bearing from housing (3—Fig. 10). Remove nut (36) and the four screws attaching steering arm (4) to housing (12), then pull steering arm from dowel pins and spindle shaft (30). Remove the six screws that attach final drive housing (12) to spindle housing (2), then separate housings.

To disassemble the final drive, remove both nuts (22), then press gear (19) from shaft (11). Shims (10) are installed to adjust mesh position of gear (19) and

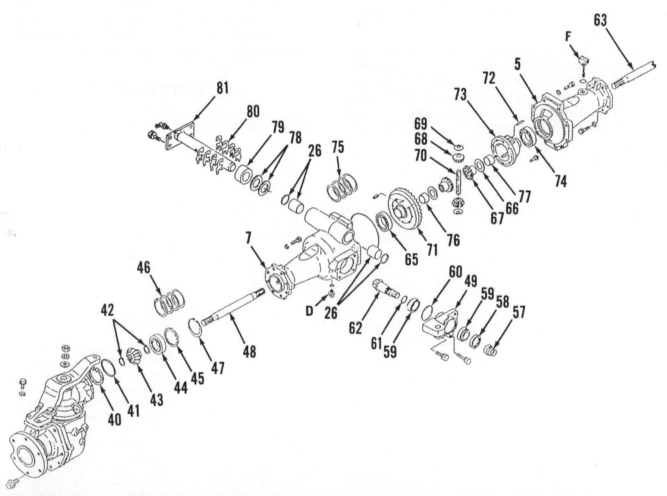

Fig. 11—Exploded view of front-wheel drive axle and differential assembly used on 870 models so equipped. Refer to Fig. 10 for spindle and final drive.

5. Right axle housing	47. Snap ring	63. Right drive shaft	72. Roll pin
7. Differential & left axle housing	48. Left drive shaft	65. Left carrier bearing	73. Differential housing
26. Pivot bushings & seal rings	49. Input shaft retainer	(same as 74)	74. Ball bearing (same as 65)
40. Snap ring	57. Spacer	66. Thrust washer	75. Shims
41. "O" ring	58. Seal	67. Side gear	76. Bushing
42. Snap rings	59. Ball bearings	68. Pinion (2 used)	77. Bushing
43. Drive gear	60. Seal ring	69. Thrust washer (2 used)	78. Shims
44. Bearing	61. "O" ring	70. Cross shaft	79. Spacer
45. Spacer	62. Input shaft and	71. Ring gear	80. Shims
46. Shims	bevel pinion		81. Pivot pin

should not be lost or damaged when removing. Shaft (11) can be pressed from bearing (17). Inspect bearings and seals for wear, roughness or other damage. Remove seal and bearings only if new parts are being installed. Wear sleeve (15) can be removed if renewal is required. Be sure to remove old gasket (13) and install new gasket. Lubricate all seals, seal surfaces and bearings before installing and assembling. Tighten first nut (22) to remove all end play, then tighten second nut to lock position. Check to be sure that shaft rotates freely after tightening locknut.

Use a soft, heavy hammer carefully to separate housing (2) from housing (3). Be careful not to lose spacer (24). Remove seal (25) if renewal is required. Remove snap ring (39) and withdraw parts (23, 27, 28, 29, 31, 37, 38 and 43). Remove key from top of shaft (30) and bump shaft from housing. Lubricate all seals, seal surfaces and bearings before installing and assembling.

Assemble by reversing disassembly procedure. Install new gasket (13) and tighten six screws connecting housings (2 and 12) to 90 N·m (66 ft.-lbs.) torque. Backlash between gears (19 and 29) should be 0.10-0.15 mm (0.004-0.006 inch) when measured at wheel attaching screw. Remove gear (19) and add shims (10) if backlash is excessive. At final assembly, use flexible sealer on gasket (13) and thread lock on screws. Tighten the four screws attaching steering arm (4) to housing (12) to 51 N·m (36 ft.-lbs.) torque and nut (36) to 187 N·m (138 ft.-lbs.) torque. Measure clearance between steering arm (4) and top of housing (3) with a feeler gauge. Steering arm should not touch hous-

ing, but if clearance is more than 0.3 mm (0.012 inch), add additional shims (34). Refer to paragraph 17 for installation of final drive and spindle to axle.

DIFFERENTIAL INPUT

Four-Wheel Drive 870 Models

19. R&R AND OVERHAUL. Refer to paragraph 16 and remove the front drive axle. Remove the four cap screws retaining cover (49—Fig. 11) to axle housing, then remove differential input cover. Shaft (62) and bearings (59) will be removed with cover. Bearings can be removed using a press or suitable pullers. Remove "O" ring (60) and clean mating surfaces.

When reassembling, install new "O" ring (60) and tighten screws attaching retainer (49) to 26 N·m (19 ft.-lbs.) torque. Refer to paragraph 16 for installing axle assembly. Install drain plugs and fill axle with lubricant as outlined in paragraph 14.

DIFFERENTIAL

Four-Wheel Drive 870 Models

20. R&R AND OVERHAUL. To remove the differential, first refer to paragraph 16 and remove axle assembly. Although not required, the final drives and steering spindles can be removed as outlined in paragraph 17. Refer to paragraph 19 and remove differential input shaft and bevel gear (62—Fig. 11).

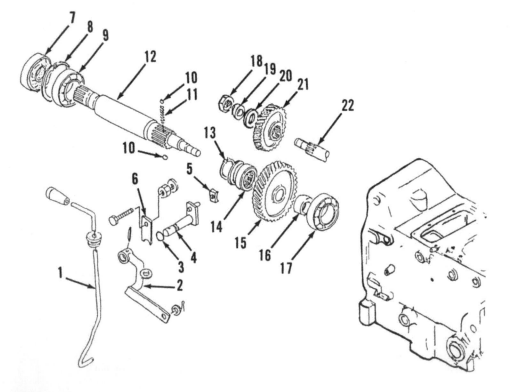

Fig. 12—Exploded view of mechanical front-wheel drive internal gears typical of 870, 970 and 1070 models so equipped. Gear (15) is driven by gear located on the front of main drive bevel pinion.

1. Shift handle	13. Snap ring
2. Lever	14. Engagement
3. "O" ring	coupling
4. Internal lever	15. Driven gear
5. Shoe	16. Spacer
6. Retainer	17. Ball bearing
7. Seal	18. Nut
8. Snap ring	19. Nut
9. Ball bearing	20. Washer
10. Detent balls	21. Gear
11. Detent spring	22. Bevel pinion
12. Shaft	shaft

Unbolt right axle housing (5) and separate from differential housing and left axle housing (7). Differential assembly can be lifted from housing. Be careful not to lose shims (75) which sets backlash between bevel pinion (62) and ring gear (71).

Use puller to remove bearings (65 and 74). Ring gear (71) is attached to differential case (73) with six socket head screws. Drive roll pin (72) from case and cross shaft (70), then drive cross shaft from case and gears (68). Roll gears (68) and thrust washers (69) from differential case, then withdraw side gears (67) and thrust washers (66).

Assembly of differential is reverse of disassembly procedure. Apply medium strength thread lock to screws retaining ring gear (71) and tighten to 40 N·m (29 ft.-lbs.) torque.

Install shims (75) that were removed when installing differential. Make sure that bearing (65) is firmly seated on ring gear (71), then install differential in housing (7). Install bevel pinion (62), bearings (59) and retainer (49), then tighten screws to 26 N·m (19 ft.-lbs.) torque. Backlash between ring gear (71) and pinion (62) should be 0.17-0.23 mm (0.007-0.009 inch) when measured at outer edge of ring gear teeth. If backlash is incorrect, remove bevel pinion (62), lift differential from housing (7), then change thickness of shims (75).

When proper thickness of shims (75) has been selected, reassembly can be completed. Install new seal ring (64) and attach left axle housing (5) to differential and right axle housing (7). Apply medium strength thread sealer to retaining screws and tighten in a cross pattern.

Refer to paragraph 17 and install both left and right final drives and steering spindles if they were removed. Refer to paragraph 16 and install the axle assembly.

FRONT-WHEEL DRIVE GEAR CASE

Four-Wheel Drive 870 Models

21. R&R AND OVERHAUL. Power for front-wheel drive is from a gear (21—Fig. 12) located on front of main drive bevel pinion shaft (22). Gear (21) drives gear (15). To remove shaft (12) and related parts, it is necessary to split clutch housing from the transmission housing as outlined in paragraph 150. Remove retainer plate (6) cap screw and pull shift arm (4) outward. Remove cover from bottom of transmission housing. Remove nuts (18 and 19), spacer (20) and drive gear (21). Pry out seal (7) and remove snap ring (8). Use a slide hammer puller to remove output shaft assembly from housing. Note that detent balls (10) will fly out of output shaft when gear (15) and engagement coupling (14) are removed.

When assembling, reverse disassembly procedure. Spring-loaded lip of front seal (8) must be toward inside. Install drive gear (21) with long end of gear hub toward rear of transmission. Tighten drive gear nuts (18 and 19) to 88 N·m (65 ft.-lbs.) torque. Reconnect transmission to clutch housing as outlined in paragraph 150. Fill transmission, rear axle center housing and hydraulic system as outlined in paragraph 178.

FOUR-WHEEL DRIVE FRONT AXLE SYSTEM (970 AND 1070 MODELS)

22. Mechanical front-wheel drive is available. Refer to the servicing instructions which follow for 970 and 1070 models.

MAINTENANCE

Four-Wheel Drive 970 And 1070 Model

23. Oil should be maintained at top mark of dipstick attached to fill plug (F—Fig. 13). Every 500 hours, drain oil from front drive axle by removing the two plugs (D) and plug (P) from housings. Removing filler plugs (F) will help housings to drain. Fill with "John Deere GL-5" or equivalent gear lubricant to correct level. Capacity is approximately 2.13 L (2.25 qt.); however, some lubricant may be trapped in housing.

Check screws retaining front wheels occasionally to make sure that they do not loosen. Tighten front wheel retaining screws to 190 N·m (140 ft.-lbs.) torque.

FRONT AXLE ASSEMBLY AND STEERING LINKAGE

Four-Wheel Drive 970 And 1070 Models

24. TIE ROD AND TOE-IN. A single tie rod (1—Fig. 14) connects left and right steering arms. Automotive type tie rod ends are not adjustable for wear and should be renewed if worn. Attach steering tie rod ends to steering arms, tighten castellated nuts to 68 N·m (50 ft.-lbs.) torque and install cotter pins.

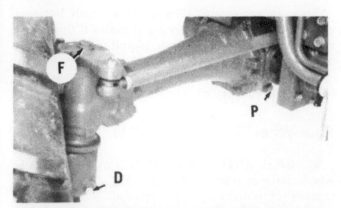

Fig. 13—View of drain plugs (D & P) and filler plug (F) for 970 and 1070 models. Dipstick is attached to filler plug.

Recommended toe-in is 3-9 mm (⅛-⅜ in.) and is adjusted by threading rod ends into tie rod to change the distance between ends. To check alignment, measure between wheel rims on centerline of axle, parallel to ground. Rotate wheels and re-measure to be sure that wheels are not bent, giving incorrect reading. Tighten rod end jam nut after wheel alignment is correctly set.

25. R&R AXLE ASSEMBLY. To remove front axle assembly, first remove any front mounted equipment and weights. Detach both ends of steering cylinder. Hoses to steering cylinder can remain attached if cylinder is moved and safely supported while detached from front axle. Raise and block front of tractor in such a way that it will not interfere with removal of axle. Support tractor behind axle and remove both front wheels. If axle unit is to be disassembled, remove plugs (D, F and P—Fig. 13) to drain oil from front axle assembly. Remove the front axle driveshaft. Support axle level with floor to prevent tipping, in such a way that it can be lowered and moved away from tractor. Remove screws retaining the axle pivot pin (60—Fig. 14), then carefully remove shims (59). Remove pivot pin (60), lower the axle assembly and roll axle safely from under tractor.

Check axle pivot bushings (30—Fig. 14) and renew if necessary. Install new bushings into bore of housing (7) using suitable bushing driver. Reverse removal procedure when assembling. Axle end play should be 0.127-1.016 mm (0.005-0.040 in.). To measure and adjust end play, push axle to the rear, then measure clearance between front of axle and plate on pivot pin with a feeler gauge. Shims (59—Fig. 14) should be added to increase clearance. Make sure that screws attaching pivot pin plate (60) are tight when measuring, but that axle does not bind. Screws should be tightened to 88 N•m (65 ft.-lbs.) torque. Attach tie rod and steering cylinder or drag link to steering arm, tighten castellated nuts to 68 N•m (50 ft.-lbs.) torque and install cotter pins. Tighten front wheel retaining

screws to 190 N•m (140 ft.-lbs.) torque. Refer to paragraph 23 and 24 for additional assembly notes.

OUTER DRIVE AXLE AND REDUCTION

Four-Wheel Drive 970 And 1070 Models

26. R&R AND OVERHAUL. Outer drive axle parts (8 through 18—Fig. 14) can be removed and serviced without additional disassembly of axle. Raise and block front of tractor in a way that is safe and will not interfere with the removal of front wheel or outer final drive. Remove plugs (D and F—Fig. 13) and drain oil from axle housing. Additional oil can be drained by removing plug (P). Remove the nine cap screws attaching housing (12—Fig. 14) to housing (2), then separate housings. Be careful not to damage or lose shims (10). Remove screws and washers (9), then press shaft (11) and gear (18) from housing. Remove bearings (8 and 17) only if renewal is required. Unbolt and remove retainers (16) to remove bearing (17) and seal (14).

Lubricate all parts and assemble by reversing disassembly procedure. Install new seal ring (13) and original shims (10) evenly distributed around housings. Tighten the nine screws attaching housings (2 and 12) to 51 N•m (39 ft.-lbs.) torque. Backlash between gears (18 and 25) should be 0.20-0.35 mm (0.008-0.014 inch) when measured at wheel attaching screw. Remove some thickness of shims (10) if backlash is excessive, then reassemble and recheck. Make sure that shims (10) are the same thickness around housings. Refer to paragraph 23 and fill axle to correct level with lubricant. Tighten front wheel retaining screws to 217 N•m (160 ft.-lbs.) torque.

FRONT FINAL DRIVE AND SPINDLE

Four-Wheel Drive 970 And 1070 Models

27. REMOVE AND REINSTALL. Raise and block front of tractor in a way that is safe and will not interfere with the removal of front wheels or final drive. Remove plugs (D and F—Fig. 13) and drain oil from axle housing. Additional oil can be drained by removing plug (P). Detach tie rod (1—Fig. 14), drag link and steering cylinder from steering arm (4), depending upon options and which final drive is being removed. Support final drive assembly safely with a floor jack, so that unit can be withdrawn, then remove the eight screws which attach housing (3) to housing (5 or 7). Carefully separate housings, then carefully move final drive away from tractor.

Remove old "O" ring (31), then lubricate new "O" ring with petroleum jelly and position on axle housing. Install the final drive and spindle, tightening retaining screws to 51 N•m (39 ft.-lbs.) torque. "Loctite" or similar thread lock should be applied to threads of retaining screws. Connect tie rod, drag link and steering cylinder to steering arm and tighten nuts to 68 N·m (50 ft.-lbs.) torque. Tighten front wheel retaining screws to 217 N·m (160 ft.-lbs.) torque. Refer to paragraph 23 and fill axle to correct level with lubricant.

28. OVERHAUL. Remove front final drive and spindle as outlined in paragraph 27, then refer to paragraph 26 for removing and servicing outer drive

axle and reduction (8 through 18—Fig. 14). Remove the four screws attaching steering arm (4) to housing (2), then bump steering arm from dowel pins and separate steering arm (4), gear case housing (2) and spindle housing (3). Shaft (35 or 43), bearing (33) and gear (32) can be withdrawn from axle housing (5 or 7) if service is required. Be careful not to damage or lose shims (34).

Remove shaft (23), gear (25) and bearing (24) from housing (3). Use puller to remove bearing (26), seal (28) and spacer (29), being careful not to lose or

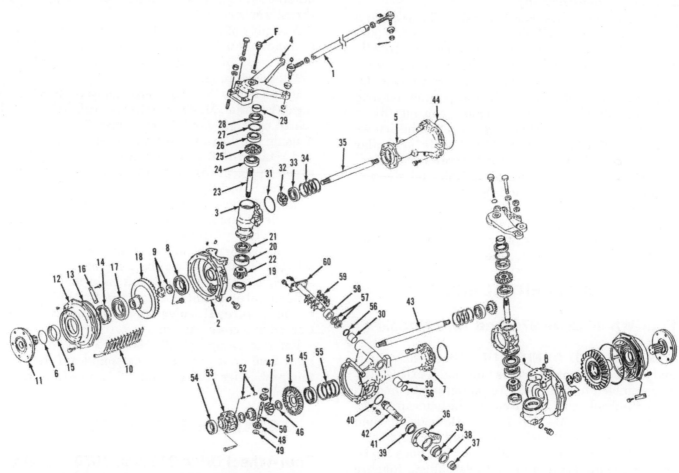

Fig. 14—Exploded view of mechanical front-wheel drive left side spindle and final drive gear case used on 970 and 1070 models with front-wheel drive axle. Parts for right side are similar.

1. Tie rod	16. Retainer plate & screws	32. Drive gear	46. Thrust washer
2. Gear case housing	17. Ball bearing	33. Ball bearing	47. Side gear
3. Spindle housing	18. Bevel gear	34. Shims	48. Pinion (2 used)
4. Left steering arm	19. Ball bearing	35. Left drive shaft	49. Thrust washer (2 used)
5. Left axle housing	20. Ball bearing	36. Input shaft retainer	50. Cross shaft
6. "O" ring	21. Oil seal	37. Spacer	51. Ring gear
7. Differential & right axle housing	22. Final drive pinion	38. Seal	52. Roll pins
8. Ball bearing	23. Spindle shaft	39. Ball bearings	53. Differential housing
9. Retainer washers & screws	24. Ball bearing	40. Seal ring	54. Ball bearing (same as 45)
10. Shims	25. Driven gear	41. "O" ring	55. Shims
11. Wheel hub & axle	26. Ball bearing	42. Input shaft and bevel pinion	56. Seal rings
12. Final drive housing	27. Shims	43. Right drive shaft	57. Shims
13. Seal ring	28. Seal	44. Seal ring	58. Spacer
14. Seal	29. Wear sleeve	45. Right carrier bearing (same as 54)	59. Shims
15. Sleeve	30. Pivot bushings		60. Pivot pin
	31. "O" ring		

damage shims (27). Remove seal (21), bearing (20), gear (22) and bearing (19) from housing (2). Lubricate all seals, seal surfaces and bearings before installing and assembling.

Assemble by reversing disassembly procedure. Install new seal ring (13) and tighten the nine screws connecting housings (2 and 12) to 51 N·m (39 ft.-lbs.) torque.

Backlash between gears (18 and 25) should be 0.20-0.35 mm (0.008-0.014 inch) when measured at wheel attaching screw. Reduce thickness of shims (10) if backlash is excessive, then reassemble and recheck. Make sure that shims (10) are the same thickness around housings.

Backlash between inner end of shaft (35 or 43) and lower gear (22) should also be 0.20-0.35 mm (0.008-0.014 inch) when measured at teeth of gear (22) while holding inner end of shaft. Reduce backlash by increasing thickness of shims (34). Lubricate new "O" ring (31) with petroleum jelly and position on axle housing. Attach assembled spindle housing (3) on axle housing (5 or 7), then tighten retaining screws to 51 N·m (39 ft.-lbs.) torque. Use "Loctite" or similar thread lock on threads of retaining screws before final assembly. Connect tie rod, drag link and steering cylinder to steering arm and tighten nuts to 68 N·m (50 ft.-lbs.) torque.

Refer to paragraph 23 and fill axle to correct level with lubricant. Tighten front wheel retaining screws to 217 N·m (160 ft.-lbs.) torque.

DIFFERENTIAL INPUT

Four-Wheel Drive 970 And 1070 Models

29. R&R AND OVERHAUL. Remove driveshaft and drain lubricant from axle housing. If axle is to be disassembled further, refer to paragraph 25 and remove the front drive axle. Remove the four cap screws retaining cover (36—Fig. 14) to axle housing, then remove differential input cover. Shaft (42) and bearings (39) will be removed with cover. Bearings can be removed using a press or suitable pullers. Remove "O" ring (40) and clean mating surfaces.

When reassembling, install new "O" ring (40) and tighten screws attaching cover (36) to 26 N·m (19 ft.-lbs.) torque. Refer to paragraph 25 for installing axle assembly. Install drain plugs and fill axle with lubricant as outlined in paragraph 23.

DIFFERENTIAL

Four-Wheel Drive 970 And 1070 Models

30. R&R AND OVERHAUL. To remove the differential, first refer to paragraph 25 and remove axle assembly. Although not required, the final drives and

steering spindles can be removed as outlined in paragraph 27. Refer to paragraph 29 and remove differential input shaft and bevel gear (42—Fig. 14).

Unbolt left axle housing (5) and separate from differential housing and right axle housing (7). Differential assembly can be lifted from housing. Be careful not to lose shims (55) which sets backlash between bevel pinion (42) and ring gear (51).

Use puller to remove bearings (45 and 54). Ring gear (51) is attached to differential case (53) with six cap screws. Drive roll pin (52) from case and cross shaft (50), then drive cross shaft from case and gears (48). Roll gears (48) and thrust washers (49) from differential case, then withdraw side gears (47) and thrust washers (46).

Assembly of differential is reverse of disassembly procedure. Apply medium strength thread lock to screws retaining ring gear (51) and tighten to 80 N·m (59 ft.-lbs.) torque.

Install shims (55) that were removed when installing differential. Make sure that bearing (45) is firmly seated on ring gear (51), then install differential in housing (7). Install bevel pinion (42), bearings (39) and retainer (36) and tighten screws. Backlash between ring gear (51) and pinion (42) should be 0.13-0.18 mm (0.005-0.007 inch) when measured at outer edge of ring gear teeth. If backlash is incorrect, remove bevel pinion (42), lift differential from housing (7), then change thickness of shims (55).

When proper thickness of shims (55) has been selected, reassembly can be completed. Install new seal ring (44) and attach left axle housing (5) to differential and right axle housing (7). Apply medium strength thread sealer to retaining screws and tighten in a cross pattern.

Refer to paragraph 27 and install both left and right final drives and steering spindles if they were removed. Refer to paragraph 25 and install the axle assembly.

FRONT-WHEEL DRIVE GEAR CASE

Four-Wheel Drive 970 And 1070 Models

31. R&R AND OVERHAUL. Power for front-wheel drive is from a gear (21—Fig. 12) located on front of main drive bevel pinion shaft (22). Gear (21) drives gear (15). To remove shaft (12) and related parts, it is necessary to split clutch housing from the transmission housing as outlined in paragraph 150. Remove retainer plate (6) cap screw and pull shift arm (4) outward. Remove cover from bottom of transmission housing. Remove nuts (18 and 19), spacer (20) and drive gear (21). Pry out seal (7) and remove snap ring (8). Use a slide hammer puller to remove output shaft assembly from housing. Note that detent balls (10) will fly out of output shaft when gear (15) and engagement coupling (14) are removed.

When assembling, reverse disassembly procedure. Spring-loaded lip of front seal (8) must be toward inside. Install drive gear (21) with long end of gear hub toward rear of transmission. Tighten drive gear nuts (18 and 19) to 88 N·m (65 ft.-lbs.) torque. Reconnect transmission to clutch housing as outlined in paragraph 150. Fill transmission, rear axle center housing and hydraulic system as outlined in paragraph 178.

MANUAL STEERING SYSTEM

Models With Manual Steering

32. Some models are equipped with a recirculating ball nut type manual steering gear similar to the type shown in Fig. 15.

33. ADJUSTMENT. Steering wheel free play should be within the range of 20-50 mm ($^{13}/_{16}$-2 inches) measured at steering wheel rim. To adjust steering play, loosen locknut (16—Fig. 15) and turn adjusting screw (12) to obtain desired free play without binding. Turn the adjusting screw clockwise to reduce free play. Tighten locknut to secure adjusting screw. The adjusting screw may not compensate for excessive wear and may cause binding if tightened too tight. Use care when adjusting.

34. R&R AND OVERHAUL. To remove the steering gear assembly, first remove nut (1—Fig. 15) and use a suitable puller to remove steering wheel. On 670 and 770 models, remove the fuel tank. On 870 and 970 models, remove the engine shields and lower dash covers. On all models, remove instrument panel and cowl. Detach drag link (20) from pitman arm (18), unbolt steering gear housing (9) from clutch housing and remove assembly.

To disassemble, drain oil from housing, remove nut (19), make sure match marks are on pitman arm (18) and steering shaft (11), then use suitable puller to remove pitman arm. Remove locknut (16) and bolts attaching side cover (14), then turn adjusting screw (12) to remove side cover. Tap sector shaft (11) out side opening. Loosen locknut (5) and remove steering col-

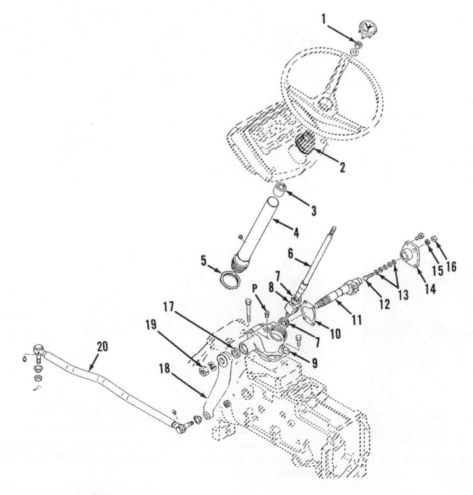

Fig. 15—Exploded view of the manual steering gear unit typical of type used on some models.

1. Nut
2. Dust seal
3. Bushing
4. Steering column
5. Locknut
6. Wormshaft
7. Bearings
8. Ball nut
9. Steering housing
10. Gasket
11. Rocker shaft
12. Adjuster screw
13. Shims
14. Cover
15. Seal washer
16. Locknut
17. Oil seal
18. Pitman arm
19. Nut
20. Drag link

umn (4). Withdraw steering shaft and ball nut assembly (6, 7 and 8). Ball nut and shaft are available only as a unit and disassembly is not recommended.

To assemble the unit, reverse the disassembly procedure while noting the following: Always install new "O" rings and oil seals. Install sector shaft (11) into case and turn fully clockwise, then install ball nut and engage it with sector gear. Install side cover using liquid sealer on gasket (10). Install steering column, then use torque wrench to test the amount of torque required to turn steering shaft (6), which should be less than 1.7 N·m (15 in.-lbs.). Turn steering column

(4) into housing to obtain desired turning torque, then tighten locknut (5) securely. Be sure to align match marks when reinstalling pitman arm and tighten retaining nut to 172 N·m (127 ft.-lbs.) torque. Fill unit to level of plug (P) with API Service Classification GL-5 gear oil.

Install steering gear assembly, tightening the steering housing to clutch housing retaining screws to 108 N·m (80 ft.-lbs.) torque and steering wheel locknut (1) to 27 N·m (20 ft.-lbs.) torque. Adjust steering wheel free play as outlined in paragraph 33.

POWER STEERING SYSTEM

All Models With Power Steering

35. The hydrostatic steering system consists of a pump, steering valve assembly, and a steering cylinder. In normal operation, pressurized hydraulic fluid is supplied by the gear type hydraulic pump. However, in the event of hydraulic failure or engine stoppage, manual steering can be accomplished by the gerotor pump in the steering valve.

FILTER AND BLEEDING

All Models With Power Steering

36. Fluid for the power steering system is contained in the transmission and rear axle center housings. The same fluid lubricates the gearbox and range transmission and is used for the main hydraulic system. Recommended fluid, "John Deere Low Viscosity HY-GARD," should be maintained at level of full mark on dipstick (D—Fig. 16A or Fig. 17). Oil fill opening is located at F—Fig. 16B or Fig. 17. New fluid and a new cartridge filter should be installed and the suction screen should be cleaned or renewed after

every 500 hours of operation. Fluid and filter should also be changed if steering unit is overhauled or if fluid is suspected of contamination. Operate the tractor to warm the hydraulic oil before draining.

To remove the cartridge filter, first lower lift arms, drain fluid from the transmission and rear axle center housing. The spin-on filter is located under the tractor on the suction (inlet) line to the power steering pump. Remove shield from around filter, unscrew

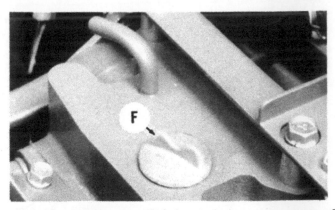

Fig. 16B—Oil filler cap (F) is located at top rear corner of rockshaft housing on 670 and 770 models.

Fig. 16A—View of dipstick (D) for transmission and hydraulic fluid typical of 670 and 770 models.

Fig. 17—View of dipstick (D) and filler plug (F) for transmission and hydraulic fluid typical of 870, 970 and 1070 models.

Fig. 18—View of cover (C) which must be removed to clean the suction screen. Remove plug (D) to drain hydraulic fluid.

filter from adapter and clean adapter housing. Lubricate seal on filter, then install and tighten new filter element.

The suction screen can be removed by unbolting cover (C—Fig. 18) and withdrawing screen. The screen can be carefully cleaned, but if damaged, install new unit.

The hydrostatic steering system is self-bleeding. When the unit has been disassembled, refill with new oil, then start engine and cycle the system several times by turning the steering wheel from lock to lock. Capacity is approximately 15 L (4 gal.) for 670 and 770 models; 21 L (5.5 gal.) for 870, 970 and 1070 models. Recheck fluid level and add fluid as required.

TROUBLESHOOTING

All Models With Power Steering

37. Some of the problems which may occur during operation of power steering and their possible causes are as follows:

A. Steering wheel hard to turn. Could be caused by:
 a. Defective power steering pump.
 b. Mechanical parts of front steering system binding.
 c. Bearings in steering column damaged.
 d. Leaking steering cylinder.
 e. Control valve spool and sleeve binding.

B. Steering wheel turns on its own. Could be caused by:
 a. Leaf springs in control valve weak or broken.

C. Steering wheel does not return to neutral position. Could be caused by:
 a. Control valve spool and sleeve jammed.
 b. Leakage between valve sleeve and housing.
 c. Dirt or metal chips between valve spool and sleeve.

D. Excessive steering wheel play. Could be caused by:
 a. Inner teeth of rotor or driveshaft teeth worn.

b. Leaf springs in control valve weak or broken.

E. Steering wheel rotates at steering cylinder stops. Could be caused by:
 a. Excessive leakage in steering cylinder.
 b. Rotor and stator excessively worn.
 c. Excessive leakage between valve spool and sleeve
 d. Excessive leakage between sleeve and housing.

F. Steering wheel "kicks" violently. Could be caused by:
 a. Incorrect adjustment between driveshaft and rotor.

G. Steering wheel responds too slowly. Could be caused by:
 a. Excessive front end weight.
 b. Not enough oil.
 c. Steering control valve worn.

H. Tractor steers in wrong direction. Could be caused by:
 a. Hoses to steering cylinder incorrectly connected.
 b. Incorrect timing of driveshaft to rotor.

SYSTEM PRESSURE AND FLOW

All Models With Power Steering

38. To check the system relief pressure, install a "T" fitting in pressure line at steering cylinder, connect a 0-21,000 kPa (0-3000 psi) test gauge to the fitting and operate engine at 1200 rpm. Turn the front wheels to one extreme against lock, and observe gauge reading. System relief pressure should be 8274-9653 kPa (1200-1400 psi).

CAUTION: When checking system relief pressure, hold the steering wheel against lock only long enough to observe pressure indicated by gauge. Pump may be damaged if steering wheel is held in this position too long or if flow is otherwise stopped.

The pressure relief valve is located in the steering control valve. Adjustment requires removal of the steering control valve. Refer to paragraph 44 for control valve removal procedure and exploded view of relief valve. Turning the relief valve spring seat in, increases the pressure against spring and increases system relief pressure. Spring seat is accessible after removing relief valve plug.

39. To determine if the steering control valve is causing steering to be hard or to wander, remove cap from center of steering wheel and turn the steering wheel fully to the right. Operate engine at high idle and attempt to turn the steering wheel nut clockwise with a torque wrench. Maintain 6.8 N·m (60 in.-lbs.) torque and observe the number of revolutions of the steering wheel (torque wrench) in one minute. Internal leakage is indicated if steering wheel turns more

PUMP

All Models With Power Steering

than five revolutions within the one minute test. Internal leakage may be occurring either in the steering cylinder, control valve or both.

40. Condition of the power steering pump can be checked by attaching a suitable flow meter to the pump outlet. Special John Deere test fitting (JDG 694) can be attached to the pump after unbolting the outlet from bottom of pump. Direct fluid back to sump by inserting return hose from flow meter into filler opening (F—Fig. 16B or Fig. 17). With engine operating at 2600 rpm, slowly close the flow meter testing valve until pressure is increased to 1600 psi, then observe the volume of flow. If flow is less than 4 gallons per minute, first check condition of the pump suction screen and the spin-on (cartridge) filter. Also, check to be sure that suction line to pump is not leaking and allowing air to enter. A worn or damaged pump will also reduce volume of flow.

41. REMOVE AND REINSTALL. The separate steering pump is attached to the front of engine timing gear cover ahead of the main hydraulic pump. Clean pump and area around pump thoroughly, then disconnect lines from pump and allow fluid to drain. Cover all openings to prevent dirt from entering pump or lines, then unbolt and remove pump from the engine timing case.

Tighten pump retaining screws to 26 N·m (226 in.-lbs.) torque. Tighten inlet and outlet flange screws to 6 N·m (53 in.-lbs.) torque. Refer to paragraph 36 for filling and bleeding.

42. OVERHAUL. Mark relative position of inlet flange (5 or 28—Fig. 19), housings (15 or 30) and covers (7 or 19), before removing retaining screws.

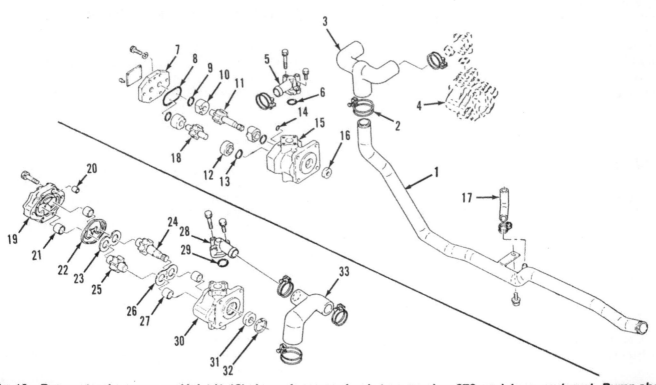

Fig. 19—Power steering pump and inlet (1-18) shown in upper view is type used on 670 models so equipped. Pump shown in lower view (19-33) is type used on other models. Power steering pump is mounted on the front of engine timing gear housing of all models. Main hydraulic system pump is shown at (4).

1. Suction line	10. Bushings	18. Pump gear	26. Wear plate
2. Clamp	11. Drive shaft & gear	19. Cover	27. Bushings
3. Manifold hose	12. Bushings	20. Dowel pins	28. Inlet flange
4. Main hydraulic pump	13. "O" rings (same as 9)	21. Bushings	29. "O" ring
5. Inlet flange	14. "O" ring	22. Packing	30. Housing
6. "O" ring	15. Pump housing	23. Wear plate	31. Oil seal
7. Cover	16. Oil seal	24. Drive shaft & gear	32. Snap ring
8. Packing	17. Hose	25. Pump gear	33. Manifold hose
9. "O" rings (4 used)			

Pump gears (11 and 18 or 24 and 25) are available only as matched set. Check condition of bearings (10 and 12 or 21 and 27), wear plates (23 and 26), gears (11 and 18 or 24 and 25) and housings (15 or 30) for wear or other damage. On pump shown in top view (7-16), mark location of each part of bearings (10 and 12) to be sure parts are reinstalled in same location.

Lubricate all parts with hydraulic fluid when assembling. On pump shown in lower view, brass side of wear plates (23 and 26) must be toward gears (24 and 25) and the oil groove must be toward inlet side of pump. On all models, small wear grooves are permissible in the larger (inlet) hole of pump bodies (15 or 30). Install seals (8 or 22) with rounded edge against rounded (radius) sides of seal groove. Marks affixed to parts before disassembling, should be aligned when reassembling. Tighten cover retaining screws to 15 N·m (133 in.-lbs.) torque. Tighten screws evenly and check for free rotation while tightening.

To run in new or rebuilt pump, install pump and run engine at idle speed for about three minutes without hydraulic pressure (only circulating hydraulic fluid), turn steering intermittently (to build pressure) for three minutes, then turn the steering wheel lock to lock for three minutes. Increase engine speed to rated rpm and recheck auxiliary and steering systems. Idle engine and check for leaks. Change filter and refill transmission oil.

POWER CYLINDER

All Models With Power Steering

43. R&R AND OVERHAUL. The cylinder is attached between axle and the right steering arm. To remove the cylinder, first detach steering hoses from cylinder, then cover all openings to prevent the entrance of dirt. Both ends of steering cylinder are attached by rod ends secured by castellated nuts and cotter pins. Remove cotter pin and nut, then bump tapered post of ball-joint from steering arm or axle.

Refer to Fig. 20 or Fig. 21. Clamp the cylinder (22) carefully in a holding fixture, clean end of cylinder, and remove snap ring (7). Bump end cap (13) into cylinder far enough to remove retaining clip (11), then pull rod (5), end cap (13), piston (19) and related parts from cylinder. Remove all old seals and install new seals before reassembling.

Lubricate and assemble steering cylinder by reversing disassembly procedure. Tighten locknut (21) to 61 N·m (45 ft.-lbs.) torque. Tighten castellated nuts retaining ball-joint to axle and to steering arm to 128 N·m (94 ft.-lbs.) torque, then install cotter pin. Attach steering hoses to proper fittings of cylinder and adjust position of steering hoses for clearance between front support, without hanging down below axle. Make sure hoses do not restrict tipping movement of axle.

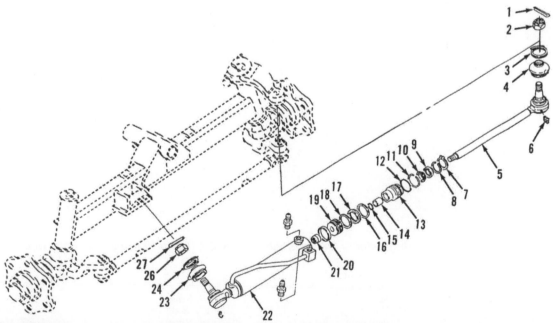

Fig. 20—Exploded view of power steering cylinder for two-wheel drive 670 and 770 models. Four-wheel drive models are similar. Refer to Fig. 21 for legend.

STEERING CONTROL VALVE

All Models With Power Steering

44. R&R AND OVERHAUL. The steering control valve is located at the base of the steering wheel shaft. To remove the control valve, first remove steering wheel retaining nut, then use suitable puller to push steering wheel from steering shaft. Remove throttle linkage, engine shields, lower dash covers, instrument panel and cowl. Remove the four banjo bolts (D and E—Fig. 22) and copper washers from each side of banjo fittings. Remove strap (B) and disconnect wiring harness at connector (F). Remove the four (wide spaced) screws (G), then lift control valve and steering column from tractor.

Thoroughly clean exterior of unit before disassembling. Remove the four screws attaching steering column (4—Fig. 23) to control valve (39), then separate valve from steering column and shaft. Remove

the seven screws (36 and 37) retaining end cover, then remove the cover (34), stator and rotor (31), wear plate (30), "O" rings (29) and driveshaft (28). Hold steering valve vertically and turn valve spool and sleeve (27) to align cross pin (26) parallel to flat side of housing. With cross pin in this position and housing in horizontal position, remove sleeve and spool (27), thrust bearing (22) and bearing races (21) from housing. Remove cross pin (26) from rotary valve, separate spool from sleeve, then remove leaf springs from spool.

The pressure relief valve (25) is not serviced as an individual unit. If necessary, the relief valve can be removed after unscrewing plugs; however, pressure must be checked and adjusted if plug is turned. Steering relief pressure is checked as outlined in paragraph 28 and adjusted by turning plug.

Clean and inspect all parts for excessive wear or other damage and renew parts as necessary. Housing

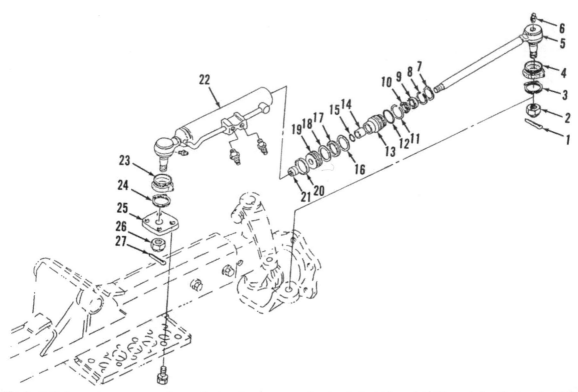

Fig. 21—Exploded view of power steering cylinder for two-wheel drive 870, 970 and 1070 models. Four-wheel drive models are similar.

1. Cotter pin	8. Snap ring	15. "O" ring	21. Locknut
2. Castellated nut	9. Scraper	16. Ring	22. Cylinder
3. Retaining ring	10. Seal	17. Seal	23. Boot
4. Boot	11. Clip	18. Ring	24. Retaining ring
5. Rod & rod end	12. Seal ring	19. Piston	25. Mounting boss
6. Grease fitting	13. Cap	20. Ring	26. Castellated nut
7. Snap ring	14. Bushing		27. Cotter pin

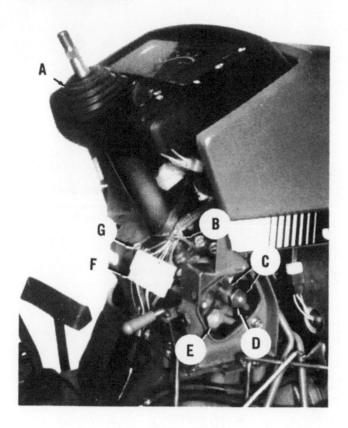

Fig. 22—View of steering column and control valve typical of all power steering equipped models. Refer to text for removal.

A. Boot
B. Tie strap
C. Washers
D. Banjo connector
E. Banjo connector
F. Rear wiring harness
G. Cap screw

(24), spool (23), sleeve (27) and relief valve (25) are not available separately. Use all new "O" rings and seals when reassembling. Lubricate all interior parts with clean hydraulic fluid.

Insert spool (23) into sleeve (27), aligning the leaf spring slots. Install leaf springs, with arch of springs together in the middle. Special tool No. KML10018-3 can be used when installing leaf springs. Insert cross pin (26) into sleeve and spool, then insert spool and sleeve assembly into housing. Install dust seal (17) with flat side toward bushing (18). Install thrust bearing (21 and 22) on spool, making sure that chamfered side of race is away from needle bearing. Install bushing and seals (17 through 20) and retain with ring (16). Turn sleeve and spool (27) until cross pin is parallel with the flat surface (ports) of housing. Mark a reference line across splined end of rotor shaft (28) parallel with the groove at other end for pin. Be sure

that valve spool pin is still aligned with the port face of the housing and install the rotor shaft as shown in Fig. 24. When correctly assembled, one lobe of the rotor will be positioned straight away from the port face (left side in Fig. 24) when pin center line is located as shown. Install one "O" ring (29—Fig. 23), wear plate (30), "O" ring (29), stator and rotor (31). Install "O" ring (32), washer (33), "O" ring (29) and end cover (34). Install the seven retaining screws (36 and 37). The steel ball (35) and the one screw with the pin attached should both be in location identified as "7" in Fig. 25. Tighten the screws in the order shown in Fig. 25, first to 11 N·m (95 in.-lbs.), then to 21 N·m (182 in.-lbs.) torque. Install relief valve assembly (25—Fig. 23) if removed. If setting of plug has been disturbed, refer to paragraph 30 for checking and adjusting relief valve pressure.

Installation is reverse of removal procedure.

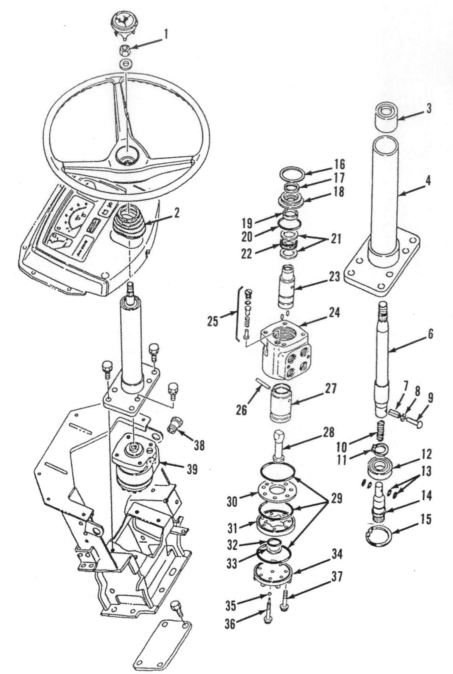

Fig. 23—Exploded view of control valve and related parts used on models with power steering.

1. Steering wheel nut
2. Boot
3. Seal
4. Steering column
5. Steering shaft
6. Steering shaft
7. Spring pin
8. Washer
9. Rivet
10. Spring
11. Snap ring
12. Ball bearing
14. Shaft
15. Snap ring
16. Retaining ring
17. Seal
18. Bushing
19. Ring seal
20. "O" ring
21. Bearing races
22. Thrust bearing
23. Valve spool
24. Housing
25. Relief valve
26. Pin
27. Valve sleeve
28. Drive shaft
29. Seal rings
30. Wear plate
31. Rotor & stator
32. "O" ring
33. Seal washer
34. End cover
35. Steel ball
36. Special screw
37. Retaining screws
38. Fitting
39. Control valve assy.

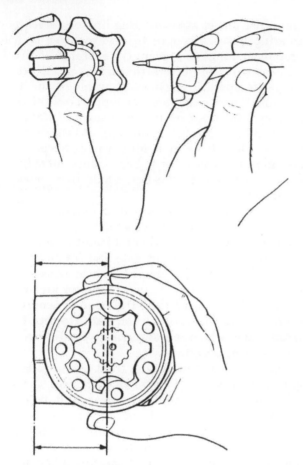

Fig. 24—The center line of pin (26—Fig. 19) should be aligned parallel with the port face of housing as shown when assembling. Center line of pin is below center line of shaft.

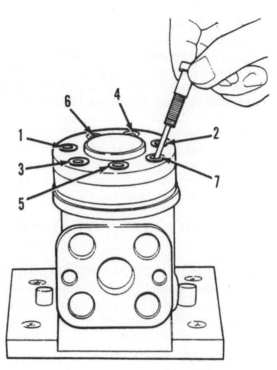

Fig. 25—The one screw with pin attached should be installed in position (7), along with steel ball (35—Fig. 23). Numbers indicate order which screws should be tightened.

ENGINE
(MODEL 670)

45. A Yanmar, three-cylinder, indirect injection, Model 3TNA72-UJX, diesel engine is used in 670 model. Refer to paragraph 75 for service to engines used in other tractor models

MAINTENANCE

670 Model

46. LUBRICATION. Recommended engine lubricant is good quality engine oil with API classification CD/SF or CD/SE. Select oil viscosity depending on ambient temperature. Recommended oil viscosity and ambient temperature ranges are as follows.

Arctic Oil . . . . −55° to 0° C
(−67° to 32° F)

SAE 5W30 . . . . −30° to +10° C
(−22° to +50° F)

SAE 10W . . . . −20° to +10° C
(−4° to +50° F)

SAE 10W30 . . . . −20° to +20° C
(−4° to +68° F)

SAE 15W30 . . . . −15° to +30° C
(+5° to +86° F)

SAE 15W40 . . . . −15° to +40° C
(+5° to +104° F)

SAE 20W40 . . . . −10° to +40° C
(+14° to +104° F)

SAE 30 . . . . 0° to +40° C
(+32° to +104° F)

SAE 40 . . . . +10° to +122° C
(+50° to +122° F)

Oil and filter should be changed after the first 50 hours of operation for new or rebuilt engine. Thereafter, oil and filter should be changed every 200 hours of normal operation. Operate tractor until oil is warm, then drain engine oil and remove oil filter. Install new filter and refill with new oil.

Capacity of engine crankcase, including filter, is 2.6 L (2.7 qt.), but oil should be maintained between marks on dipstick at all times.

47. AIR FILTER. The air filter should be cleaned after every 200 hours of operation or more frequently if operating in extremely dusty conditions. Remove dust from filter element by tapping lightly with your hand. Low pressure air (less that 210 kPa or 30 psi) can be directed through filter from the inside, toward outside, to blow dust from filter. Be careful not to damage filter element while attempting to clean. A new element should be installed if condition of old element is questionable. If special air cleaner solution and water are used to wash air cleaner, make sure element is dry before installing.

NOTE: Do not wash filter with any petroleum solvent.

48. WATER PUMP/FAN/ALTERNATOR BELT. Fan belt tension should be maintained so belt will deflect approximately 13 mm (½ inch) when moderate thumb pressure is applied to belt midway between pump and alternator pulleys. Reposition alternator to adjust belt tension.

49. FUEL FILTER. A renewable paper filter element type fuel filter is located between fuel tank and fuel injection pump. Check the clear filter sediment bowl daily for evidence of water or sediment. Bowl should be drained and new filter installed at least every 200 hours of operation or if filter appears dirty. Air must be bled from filter and fuel system as outlined in paragraph 111 each time filter is removed.

R & R ENGINE ASSEMBLY

670 Model

50. To remove engine from tractor, first remove any weights or front mounted equipment. Remove dash lower cover, then unbolt and remove hood. Remove engine side panels, disconnect battery and remove electric starter. Close fuel shut-off at fuel filter, disconnect fuel line at filters, drain fuel from tank, then cover all openings to prevent the entrance of dirt. Disconnect fuel gauge wires and fuel return hose at top of tank. Remove fuel tank mounting screws, mounting brackets and other equipment which would retain tank, then lift tank from tractor. Disconnect throttle control rod. Drain coolant, disconnect or remove radiator hoses, remove muffler and remove air cleaner assembly, then cover all openings. Lower rockshaft lift arms, disconnect or remove hydraulic lines which would interfere with the removal of en-

gine. Disconnect steering drag link or detach power steering hoses. Be sure to always cover hydraulic system openings to prevent the entrance of dirt. Disconnect wire connectors to alternator, oil pressure gauge, temperature gauge, glow plugs and to fuel shut-off. Remove the radiator support rod and coolant recovery tank (located at front of engine). Remove the air shield and the fuel tank support bracket (behind engine). Detach clamp holding wiring harness to top of engine, then move wiring and cables out of the way. Remove the front axle driveshaft from models so equipped.

Remove rocker arm cover from engine, attach two lifting eyes to engine, then attach hoist to lifting eyes. Wedge between front axle and frame to prevent tipping, then stabilize front axle and frame with floor jack under front of frame. Block rear wheels to prevent rolling and support rear of tractor under clutch housing. **Make sure that the rear of tractor, the engine and the front axle, radiator and front frame are all three separately, safely and securely supported.**

Remove screws attaching front frame rails to engine (six screws are located on each side), then carefully roll front frame, radiator and front axle assembly forward away from tractor.

Remove the eight screws attaching engine to clutch housing, shift transmission to neutral, then move the supported engine away from the clutch housing.

Reassemble by reversing removal procedure. Coat splines of clutch shaft lightly with a multipurpose grease before assembling. It may be necessary to turn flywheel to align clutch splines. Tighten the eight screws attaching engine to clutch housing to 54 N·m (40 ft.-lbs.) torque. Apply medium strength "Loctite" or equivalent to threads of the twelve screws attaching front frame to sides of engine before tightening to 90 N·m (65 ft.-lbs.) torque. Adjust hood latches to provide 10 mm (0.394 inch) hood clearance. Adjust by turning screws after loosening latch clamp screws. Refer to paragraph 111 for bleeding the fuel system.

CYLINDER HEAD

670 Model

51. REMOVE AND REINSTALL. To remove cylinder head, drain coolant and engine oil. Remove water pump, rocker arm cover, rocker arm assembly, valve caps and push rods. Remove fuel injection lines and fuel return lines. The inlet manifold, exhaust manifold, glow plugs and injection nozzles do not need to be removed to remove cylinder head. However, if required for other service, these parts should be removed for safety before removing head. Remove retaining screws then lift cylinder head from engine.

Refer to appropriate paragraphs for servicing valves, guides and springs. Clean cylinder block and

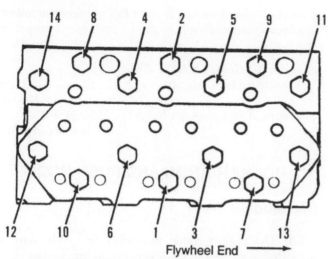

Fig. 26—Cylinder head retaining screws should be tightened in at least three stages in the sequence shown. Refer to text for recommended torque values for each stage.

cylinder head gasket surfaces and inspect cylinder head for cracks, distortion or other defects. If alignment pins were removed with cylinder head, remove pins and install in cylinder block. Check head gasket surface with a straight edge and feeler gauge. Recondition or renew cylinder head if distortion exceeds 0.15 mm (0.006 inch). Cylinder head thickness should not be reduced more than 0.20 mm (0.008 inch).

Place new cylinder head gasket over alignment pins and make sure that oil passage in gasket is positioned over passage in block. Lubricate cylinder head bolts with engine oil and install bolts, tightening in three steps in the sequence shown in Fig. 26. Refer to the following for correct three step torques.

Engine Model 3TNA72UJX

First torque	18.9 N•m
	(13.5 ft.-lbs.)
Second torque	37.2 N•m
	(27.4 ft.-lbs.)
Third (final) torque	61 N•m
	(45 ft.-lbs.)

Tighten stud nuts retaining rocker arm assembly to 26 N•m (19 ft.-lbs.) torque. Tighten rocker arm cover retaining nuts to 18 N•m (13 ft.-lbs.) torque. Tighten injector nozzle to cylinder head to 50 N•m (37 ft.-lbs.) torque and nozzle leak-off line nuts to 40 N•m (30 ft.-lbs.) torque. Tighten inlet manifold retaining screws to 11 N•m (96 in.-lbs.) torque. Tighten exhaust manifold and water pump retaining nuts and screws to 26 N•m (19 ft.-lbs.) torque.

VALVE CLEARANCE

670 Model

52. Valve clearance should be adjusted with engine cold. Specified clearance is 0.2 mm (0.008 inch) for both the inlet and exhaust valves. Adjustment is make by loosening locknut and turning adjusting screw at push rod end of rocker arm.

To adjust valve clearance, turn crankshaft in normal direction of rotation so that rear (number 1) piston is at top dead center of compression stroke. Adjust clearance of inlet and exhaust valves for rear cylinder. Turn crankshaft 240° until next piston in firing order (front) is at top dead center and adjust clearance for both valves of that cylinder. Turn crankshaft another 240° until last piston in firing order (center) is at top dead center, then adjust valve clearance for both valves of the last cylinder.

VALVES, GUIDES AND SEATS

670 Model

53. Inlet and exhaust valves seat directly in cylinder head; however, seat inserts (3 and 5—Fig. 27) may be available for service. Inlet valve face and seat angle is 30° and exhaust valve face and seat angle is 45°. Refer to the following recommended valve, guide and seat specifications.

Engine Model 3TNA72UJX

Inlet valve seat width —	
Recommended	1.44 mm
	(0.057 in.)
Maximum limit	1.98 mm
	(0.078 in.)
Exhaust valve seat width —	
Recommended	1.77 mm
	(0.070 in.)
Maximum limit	2.27 mm
	(0.089 in.)
Inlet valve seat recession —	
Recommended	0.50 mm
	(0.020 in.)
Exhaust valve seat recession —	
Recommended	0.85 mm
	(0.034 in.)
Inlet and exhaust valve stem OD	6.90 mm
	(0.272 in.)
Inlet and exhaust valve guide ID	7.08 mm
	(0.279 in.)
Valve guide height.	9.00 mm
	(0.354 in.)

Valve stem to guide clearance should not exceed 0.15 mm (0.006 inch); however, guides can be knurled to reduce clearance if less than 0.20 mm (0.008 inch).

Install new guide if clearance exceeds 0.20 mm (0.008 inch) with new valve. Press new guide (6) into cylinder head until height of guide above surface of cylinder head is 9.0 mm (0.354 inch). Valve seat width can be narrowed using 60° and 30° stones for exhaust valve seats; 45° and 15° stones for inlet valve seats.

VALVE SPRINGS

670 Model

54. Valve springs (8—Fig. 27) are interchangeable for inlet and exhaust valves. Renew springs which are

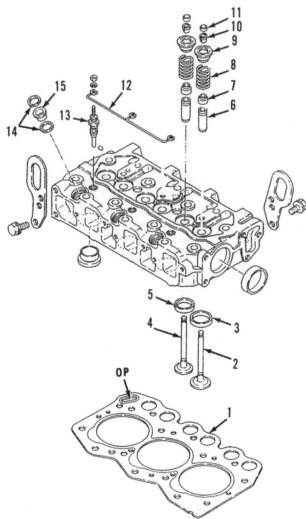

distorted, discolored by heat or fail to meet the following test specifications.

Engine Model 3TNA72UJX

Valve spring free length . . . . 36.9 mm
(1.453 in.)
Pressure at compressed height . . . 299 N @ 22.5 mm
(67 lbs. @ 0.886 in.)

ROCKER ARMS AND PUSH RODS

670 Models

55. The rocker arms assembly (Fig. 28) can be unbolted and removed after removing rocker arm cover. Push rods can be withdrawn after removing the rocker arms. Rocker arms are not interchangeable and must be assembled to position ends correctly over valve and push rod. All valve train components should be identified as removed so they can be reinstalled in original locations if reused. Install new component if old part does not meet standard listed in the following specifications.

Engine Model 3TNA72UJX

Rocker shaft OD, Min. . . . . 11.96 mm
(0.471 in.)
Shaft support ID, Max. . . . . 12.09 mm
(0.476 in.)
Rocker arm ID, Max. . . . . 12.09 mm
(0.476 in.)
Rocker arm to shaft clearance, Max. . . . . . 0.13 mm
(0.005 in.)
Push rod runout, Max. . . . . 0.30 mm
(0.012 in.)
Push rod length, Min. . . . . 141.0 mm
(5.55 in.)

Fig. 27—Exploded view of cylinder head typical of 670 model. Valve stem seal (7) is different for inlet and exhaust valves. Oil passage in cylinder head gasket (1) is located at (OP).

1. Gasket
2. Inlet valve
3. Valve seat
4. Exhaust valve
5. Valve seat
6. Valve guide
7. Valve stem seal
8. Valve spring
9. Spring seat
10. Split retainer
11. Valve cap
12. Glow plug connector
13. Glow plug
14. Gaskets
15. Seal

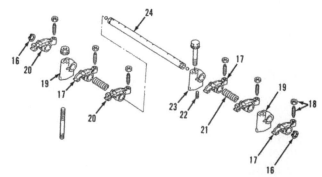

Fig. 28—Exploded view of rocker arm and shaft assembly typical of 670 model.

16. Snap ring
17. Inlet rocker arm
18. Adjuster screw & locknut
19. Support
20. Exhaust rocker arm
21. Spacer spring
22. Set screw
23. Center support
24. Shaft

Align set screw (22) in center support (23—Fig. 28) with hole in shaft (24), then tighten set screw. Assemble rocker arms, springs and remaining supports, then install snap rings (16) at ends of shaft. Tighten rocker arm assembly mounting screw and stud nuts to 26 N·m (230 in.-lbs.) torque. Adjust valve clearance as outlined in paragraph 52. Tighten rocker arm cover retaining nuts to 18 N·m (160 in.-lbs.) torque.

CAM FOLLOWERS

670 Model

56. The barrel type cam followers can be removed from engine after removing cylinder head as outlined in paragraph 51. Cam followers must be identified as they are removed so that they can be reinstalled in same bores.

Minimum outside diameter of cam follower is 20.93 mm (0.824 inch). Maximum inside diameter of cam follower bore in cylinder block is 21.05 mm (0.829 inch). Cam follower to bore clearance should not exceed 0.15 mm (0.006 inch).

TIMING GEARS

670 Model

57. The timing gear train consists of crankshaft gear (1—Fig. 29), idler gear (2), camshaft gear (3) and fuel injection pump camshaft gear (4). Timing gear marks are visible after removing timing gear cover as outlined in paragraph 62. Valves and injection pump are properly timed to crankshaft when marks on gears are aligned as shown in Fig. 29. Because of the number of teeth on idler gear, timing marks will not align every other revolution of the crankshaft. Alignment can be checked by turning crankshaft until keyway is vertical, then checking location of mark on camshaft gear. If mark on camshaft gear is away from idler gear, turn crankshaft one complete turn until keyway is again vertical. Remove idler gear, then reinstall idler gear with marks on all gears aligned as shown in Fig. 29.

NOTE: Before removing any gears, first refer to paragraph 55 and remove rocker arm assembly to avoid possible damage to piston or valve train. Damage could result if either camshaft or crankshaft is turned independently from the other unless rocker arms are removed.

Backlash between timing gears (1, 2, 3 or 4) and any one of the other timing gears should be less than 0.20 mm (0.008 inch). Backlash between oil pump gear (5) and crankshaft gear (1) should be less than 0.25 mm (0.010 inch).

Fig. 29—View of timing gears typical of 670 model.

1. Crankshaft gear
2. Idler gear
3. Camshaft gear
4. Injection pump camshaft gear
5. Oil pump gear

All timing gears (1, 2, 3 and 4) have a single timing mark at (T1, T2 and T3). Marks (T1 and T3) on idler gear are located on teeth and mark (T2) on idler gear is located in tooth valley.

58. IDLER GEAR AND SHAFT. To inspect or remove the idler gear (2—Fig. 29) and shaft (12—Fig. 30), first remove timing gear cover as outlined in paragraph 62. Check backlash between idler gear and other timing gears before removing the gear. Backlash between idler gear and any other gear (1, 3 or 4—Fig. 29) should be less than 0.20 mm (0.008 inch).

NOTE: Before removing any gears, first refer to paragraph 55 and remove rocker arm assembly to avoid possible damage to piston or valve train. Damage could result if either camshaft or crankshaft is turned independently from the other unless rocker arms are removed.

Remove snap ring (15—Fig. 30) from end of idler shaft, then withdraw gear from idler shaft. Idler shaft (12) can be unbolted from front face of cylinder block. Inside diameter of bushing (13) should be 20.00-20.021 mm (0.786-0.788 inch), but should not exceed 20.08 mm (0.761 inch). Outside diameter of shaft should be 19.959-19.980 mm (0.786-0.787 inch), but should not be less than 19.93 mm (0.785 inch). Renew shaft and/or bushing inside gear if clearance exceeds 0.15 mm (0.006 inch). Oil hole in bushing must be aligned with hole in idler gear. Inspect gear teeth for wear or scoring.

To install idler gear, position crankshaft with rear (number 1) piston at top dead center. Turn camshaft gear (3—Fig. 29) and fuel injection pump drive gear (4) so that timing marks point to center of the idler gear shaft. Install idler gear (2) so that all timing marks are aligned as shown in Fig. 29, then install idler gear retaining snap ring.

59. CAMSHAFT GEAR. The camshaft must be removed as outlined in paragraph 64 before gear (3—Fig. 30) can be pressed from shaft. Check camshaft end play and backlash between camshaft gear (3—Fig. 29) and idler gear (2) before removing camshaft. Backlash should be 0.04-0.12 mm (0.0016-0.0047 inch), but should not exceed 0.20 mm (0.008 inch). Camshaft end play should be 0.05-0.15 mm

(0.002-0.006 inch), but should not exceed 0.4 mm (0.016 inch). If end play is excessive, install new thrust plate (20—Fig. 30).

Remove rocker arm assembly as outlined in paragraph 55 and idler gear as outlined in paragraph 58. Damage could result if either camshaft or crankshaft is turned independently from the other unless rocker arms are removed. Remove camshaft as outlined in paragraph 64. Gear can be pressed from camshaft to renew thrust plate (20) or gear (3).

When reinstalling camshaft in all models, tighten screws retaining thrust plate to front of engine to 11 N•m (96 in.-lbs.) torque. Refer to paragraph 55 for installation of rocker arm assembly, paragraph 62 for installation of timing gear cover and to paragraph 52 for adjusting valve clearance.

60. CRANKSHAFT GEAR. Crankshaft gear (1—Fig. 29) is a press fit on crankshaft and should not be removed unless a new gear is to be installed. Backlash between crankshaft gear (1) and idler gear (2) should be 0.11-0.19 mm (0.0043-0.0075 inch), but

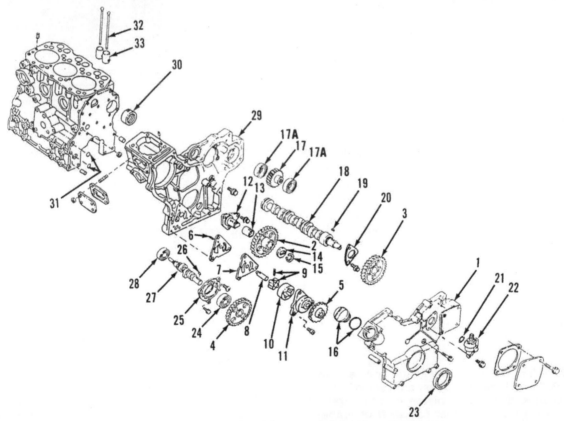

Fig. 30—Exploded view of timing gear cover and timing gear housing typical of 670 model.

1. Timing gear cover	10. Outer rotor	18. Valve camshaft	26. Drive key
2. Idler gear	11. Pump housing	19. Drive key	27. Injection pump camshaft
3. Camshaft drive gear	12. Idler shaft	20. Thrust plate	28. Ball bearing
4. Injection pump drive gear	13. Idler gear bushing	21. "O" ring	29. Timing gear housing
5. Oil pump drive gear	14. Thrust washer	22. Tachometer gear unit	30. Camshaft bearing
6. Gasket	15. Snap ring	23. Crankshaft front seal	31. "O" rings
7. Oil pump cover	16. Filler cap and gasket	24. Ball bearing	32. Cam follower
8. Pump shaft	17. Hydraulic pump drive	25. Retainer	33. Push rod
9. Inner rotor & drive pin	17A. Ball bearings		

should not exceed 0.20 mm (0.008 inch). Backlash between crankshaft gear (1) and oil pump gear (5) should be 0.11-0.19 mm (0.0043-0.0075 inch), but should not exceed 0.25 mm (0.010 inch).

To remove crankshaft gear (1), the crankshaft should be removed and gear pressed from shaft.

61. INJECTION PUMP DRIVE GEAR. Refer to paragraph 117 for removal or other service procedures to the injection pump drive gear (4—Fig. 30).

TIMING GEAR COVER

670 Model

62. Crankshaft front oil seal (23—Fig. 30) can be removed and new seal installed after removing crankshaft pulley. Seal should be driven into cover until flush with front of cover. Lubricate seal before installing crankshaft pulley.

To remove timing gear cover (1), first refer to paragraph 50 and separate front end and frame rails from engine. Remove fan and alternator. Remove retaining screw from center of crankshaft pulley. Notice that screw in center of crankshaft pulley should have "Loctite" or equivalent on threads and may be difficult to loosen. Use an appropriate puller and remove crankshaft pulley from crankshaft. Unbolt and remove timing gear cover from front of engine.

Clean mating surfaces of timing gear cover and timing gear housing, then apply an even bead of sealer to mating surface. A gasket is not used between cover and housing. Install cover and tighten retaining screws to 9 N·m (78 in.-lbs.) torque. The crankshaft pulley of some models is equipped with a pin which must engage a hole in crankshaft gear. Coat threads of center screw with medium strength "Loctite" or equivalent and tighten to 115 N·m (85 ft.-lbs.) torque. Remainder of assembly is reverse of disassembly.

TIMING GEAR HOUSING

670 Model

63. To remove timing gear housing (29—Fig. 30), first remove engine as outlined in paragraph 50, rocker arm cover and rocker arm assembly as outlined in paragraph 55, the timing gear cover (1) as outlined in paragraph 62, camshaft and drive gear (3 and 18) as outlined in paragraph 64, and injection pump camshaft and drive gear (4 and 27) as outlined in paragraph 117. Unbolt and remove oil pump (5 through 11) from timing gear housing and oil sump from bottom of engine, then unbolt timing gear housing (29) from front of cylinder block.

When reinstalling timing gear housing, reverse removal procedure. Clean and dry sealing surfaces of timing gear housing and cylinder block, then apply bead of RTV sealer on block mating surface of timing gear housing. Install timing gear housing and tighten retaining screws to 9 N·m (78 in.-lbs.) torque.

CAMSHAFT

670 Model

64. To remove the engine camshaft, first remove rocker arm cover and rocker arm assembly as outlined in paragraph 55 and timing gear cover as outlined in paragraph 62. Check camshaft end play and backlash between camshaft gear (3—Fig. 29) and idler gear (2) before removing camshaft. Backlash should be 0.04-0.12 mm (0.0016-0.0047 inch), but should not exceed 0.20 mm (0.008 inch). Camshaft end play should be 0.05-0.15 mm (0.002-0.006 inch), but should not exceed 0.4 mm (0.16 inch). If end play is excessive, install new thrust plate (20—Fig. 30). Remove and invert engine to hold cam followers (32) away from camshaft (18).

> **NOTE: It may not be necessary to turn engine up-side-down if magnets are available to hold all of the cam followers away from camshaft.**

Remove camshaft thrust plate retaining screws through the web holes of camshaft gear (3), then pull camshaft (18) out of engine block bores. Gear can be pressed from camshaft to renew thrust plate (20) or gear. Flywheel must be removed to remove plug from rear of camshaft bore in cylinder block. Refer to the following specifications.

Engine Model 3TNA72UJX
Camshaft journal OD —

Front	39.94-39.96 mm
	(1.572-1.573 in.)
Wear limit	39.85 mm
	(1.569 in.)
Intermediate	39.910-39.935 mm
	(1.571-1.572 in.)
Wear limit	39.85 mm
	(1.569 in.)
Rear	39.94-39.96 mm
	(1.572-1.573 in.)
Wear limit	39.85 mm
	(1.569 in.)
Camshaft front bearing ID	40.0-40.065 mm
	(1.575-1.577 in.)
Wear limit	40.10 mm
	(1.579 in.)
Camshaft end play	0.05-0.15 mm
	(0.002-0.006 in.)
Wear limit	0.4 mm
	(0.16 in.)

Cam lobe height. . . . . 33.95-34.05 mm
(1.337-1.341 in.)
Wear limit. 33.75 mm
(1.329 in.)

Be sure to align oil hole in bushing with passage in cylinder block if new bushing (30—Fig. 30) is installed.

When reinstalling camshaft in all models, tighten screws retaining thrust plate to front of engine to 11 N·m (96 in.-lbs.) torque. Refer to paragraph 55 for installation of rocker arm assembly, paragraph 62 for installation of timing gear cover and to paragraph 52 for adjusting valve clearance.

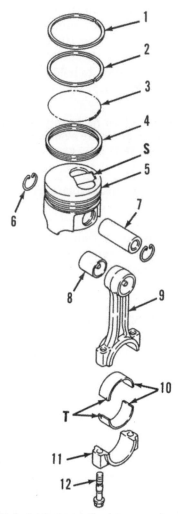

Fig. 31—Exploded view of piston and connecting rod assembly. Assemble piston to connecting rod so that slot (S) in top of piston will be on injector side of engine and slots in connecting rod and cap for bearing insert tangs (T) will be on opposite (left) side as shown.

1. Top compression ring
2. Second compression ring
3. Oil ring expander
4. Oil control ring
5. Piston
6. Retaining ring
7. Piston pin
8. Pin bushing
9. Connecting rod
10. Crankpin bearing insert
11. Connecting rod cap
12. Retaining screw

ROD AND PISTON UNITS

670 Model

65. Pistons and connecting rod units are removed from above after removing oil sump and cylinder head. Unbolt and remove oil pump suction tube and be sure that ridge (if present) is removed from top of cylinder bore before removing pistons. If not already marked, stamp cylinder number on piston, rod and cap before removing. Number "1" cylinder is at rear and connecting rods should be marked on injection pump (right) side. Keep rod bearing cap with matching connecting rod.

Piston must be assembled to connecting rod with recessed slot (S—Fig. 31) in top of piston opposite slot in connecting rod for bearing insert tang (T). Screws attaching connecting rod cap to rod should not be reused. Always install **new** connecting rod screws (12).

Lubricate cylinder, crankpin and ring compressor with clean engine oil. Space piston ring end gaps evenly around piston, but not aligned with piston pin, then compress piston rings with suitable compressor tool. Install piston and rod units in cylinder block with recessed slot (S) in top of piston on same side as fuel injection pump (right). Make sure that bore in connecting rod and cap is completely clean and install bearing inserts with tangs engaging slot in rod and cap. Make sure that inserts are firmly seated, lubricate crankpin and bearing inserts, then seat connecting rod with insert against crankpin. Install connecting rod cap with insert firmly seated and lubricated. Tangs on both bearing inserts should be on left (camshaft) side of engine. Tighten connecting rod cap retaining screws to 23 N·m (200 in.-lbs.) torque. Install new seal and oil pump pickup tube, then tighten retaining screws to 11 N·m (96 in.-lbs.) torque. Clean and dry sealing surface of oil sump and cylinder block, then coat sealing surface with RTV sealer and install oil sump. Tighten screws retaining oil sump to cylinder block to 11 N·m (96 in.-lbs.) torque and screws retaining oil sump to timing gear housing to 9 N·m (78 in.-lbs.) torque. Remainder of assembly is reverse of disassembly.

Clearance between flat top of piston and cylinder head can be measured using 10 mm (0.4 inch) long sections of 1.5 mm (0.06 inch) diameter soft lead wire positioned on top of the piston. Turn crankshaft one complete revolution, then remove lead wire and measure thickness of the flattened wire. Clearance should be 0.61-0.79 mm (0.024-0.031 inch).

PISTON, RINGS AND CYLINDER

670 Model

66. Pistons are fitted with two compression rings and one oil control ring. Piston pin is full floating and retained by a snap ring at each end of pin bore in piston. If necessary to separate piston from connecting rod, remove snap rings and push pin from piston and connecting rod. Service pistons are available in standard size and 0.25 mm (0.010 inch) oversize.

Inspect all piston skirts for scoring, cracks and excessive wear. Measure piston skirt at right angles to pin bore 8.0 mm (0.315 inch) from bottom of skirt. Minimum skirt measurement for standard size piston is 71.81 mm (2.827 inches).

Piston pin bore in piston should not exceed 21.02 mm (0.828 inch). Clearance between piston pin and bore in piston should not exceed 0.05 mm (0.002 inch).

Standard cylinder bore diameter is 72.00-72.03 mm (2.835-2.836 inches) and should not exceed 72.20 mm (2.843 inches). Piston to cylinder bore clearance should not exceed 0.28 mm (0.011 inch). Cylinders can be bored and fitted with 0.25 mm (0.10 inch) oversize pistons.

Cylinder bore should be deglazed using 180 grit stone and crosshatch pattern should intersect at 30-40°. Wash abrasive from cylinder walls using warm soapy water. Continue cleaning until white cleaning cloths show no discoloration. Make sure all soap is rinsed from cylinders, dry cylinder block thoroughly, then lubricate with clean engine oil.

> **NOTE: Use only mild soap to clean abrasive from cylinders. Do not use commercial solvents, gasoline or kerosene to clean cylinders.**

End gap of piston rings in cylinder should not exceed 1.50 mm (0.059 inch). Rings should be inserted squarely in cylinder using piston to push ring approximately 30 mm (1.181 inch) into cylinder.

Piston ring side clearance in grooves should not exceed 0.20 mm (0.008 inch) for all rings.

When installing piston rings, install oil ring expander in groove, followed by the oil ring. End gap of oil ring should be 180° from ends of expander. Install compression ring with inside chamfer or identification mark toward top of piston. Space end gaps of top, second and oil control rings 120° apart before installing in cylinder bore.

PISTON PIN

670 Model

67. Piston pin bore in piston should not exceed 21.02 mm (0.828 inch) and clearance between piston pin and bore in piston should not exceed 0.05 mm (0.002 inch).

Piston pin diameter should be measured at each end and in center. Measure pin diameter at each of the three locations and again 90° from each of the original locations. Piston pin diameter should not be less than 20.98 mm (0.826 inch).

Inside diameter of pin bushing in top of connecting rod should not exceed 21.10 mm (0.831 inch). Clearance between piston pin and bore of pin bushing in connecting rod should not exceed 0.11 mm (0.004 inch).

CONNECTING ROD AND BEARINGS

670 Model

68. Connecting rod can be unbolted from crankshaft crankpin and new bearing inserts installed from below after removing oil sump. Refer to paragraph 65 for removing connecting rods. Connecting rod should not be out-of-parallel or twisted more than 0.08 mm (0.003 inch).

Inside diameter of pin bushing in top of connecting rod should be 21.025-21.038 mm (0.8278-0.8282 inch), but should not exceed 21.10 mm (0.831 inch). Clearance between piston pin and bushing bore in connecting rod should be 0.025-0.047 mm (0.0010-0.0019 inch), but should not exceed 0.11 mm (0.004 inch).

Connecting rod bearing clearance on crankshaft crankpin should be 0.020-0.072 mm (0.0008-0.0028 inch) and wear limit should not exceed 0.15 mm (0.006 inch). Connecting rod side play on crankpin should be 0.2-0.4 mm (0.0079-0.0157 inch), but should not exceed 0.55 mm (0.0217 inch).

Crankpin standard diameter is 39.97-39.98 mm (1.5736-1.574 inch). Undersized bearing inserts, which require crankshaft journals to be resized and polished, are available for service. Finished undersize journals should provide standard clearance with undersized bearings.

Refer to paragraph 65 if connecting rod and piston units are removed from engine. Make sure that bore in connecting rod and cap is completely clean. Install bearing inserts with tangs engaging slot in rod and cap. Make sure that inserts are firmly seated, lubricate crankpin and bearing inserts, then seat connecting rod with insert against crankpin. Install connecting rod cap with insert firmly seated and lubricated. Tangs (T—Fig. 31) on both halves of bearing should be on left (camshaft) side of engine. Tighten connecting rod cap retaining screws to 23 N·m (200 in.-lbs.) torque. Install new seal and oil pump pickup tube, then tighten retaining screws to 11 N·m (96 in.-lbs.) torque. Clean and dry sealing surface of oil sump and cylinder block, then coat sealing surface with RTV sealer and install oil sump.

Tighten screws retaining oil sump to cylinder block to 11 N·m (96 in.-lbs.) torque and screws retaining oil sump to timing gear housing to 9 N·m (78 in.-lbs.) torque.

CRANKSHAFT AND MAIN BEARINGS

670 Model

69. The crankshaft is supported in four main bearings. Crankshaft end play is controlled by insert type thrust bearing (8 and 10—Fig. 32) on sides of rear (number 1) main bearing journal.

To remove crankshaft (5), remove engine from tractor as outlined in paragraph 50, rocker arm cover and rocker arm assembly as outlined in paragraph 55, timing gear cover as outlined in paragraph 62, camshaft as outlined in paragraph 64, injection pump drive gear and camshaft as outlined in paragraph 117 and timing gear housing as outlined in paragraph 63. Remove oil sump as outlined in paragraph 73. Remove flywheel (18), then remove crankshaft rear oil seal retainer housing (13). Invert engine, then refer to paragraph 65 and detach connecting rods from crankshaft. Mark all main bearing caps for correct installation, unbolt and remove crankshaft main bearing caps, then lift crankshaft from engine. The rear main bearing is number "1."

Check all main bearing journals, thrust surfaces of rear journal and connecting rod crankpin journals for scoring or excessive wear. Main bearings and rod bearings are available in standard size and 0.25 mm (0.010 inch) undersize. Refer to the following specifications.

Engine Model 3TNA72UJX

Crankpin standard OD—
Desired . . . . 39.97-39.98 mm (1.5736-1.574 in.)
Wear limit . . . . 39.92 mm (1.572 in.)

Crankpin to rod bearing clearance —
Desired . . . . 0.020-0.072 mm (0.0008-0.0028 in.)
Wear limit . . . . 0.15 mm (0.006 in.)

Main journal standard OD —
Desired . . . . 43.97-43.98 mm (1.731-1.732 in.)
Wear limit . . . . 43.92 mm (1.729 in.)

Main journal to bearing clearance —
Desired . . . . 0.020-0.072 mm (0.0008-0.0028 in.)
Wear limit . . . . 0.15 mm (0.006 in.)

Crankshaft end play —
Desired . . . . 0.090-0.271 mm (0.004-0.011 in.)
Wear limit . . . . 0.33 mm (0.013 in.)

Main cap torque . . . . 79 N·m (58 ft.-lbs.)

Rear oil seal case torque —
Seal to block . . . . 11 N·m (96 in.-lbs.)
Oil sump to seal case . . . . 9 N·m (78 in.-lbs.)

If crankshaft gear (1—Fig. 32) was removed, heat gear to approximately 150° C (300° F) before pressing

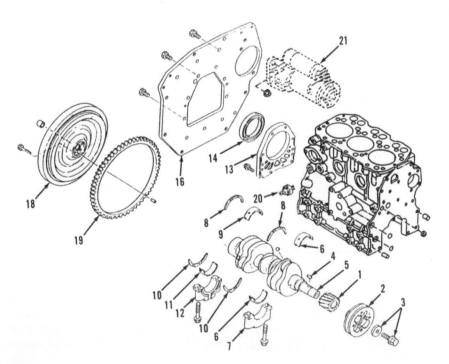

Fig. 32—View of crankshaft, rear seal and flywheel.

1. Crankshaft gear
2. Crankshaft pulley
3. Retaining screw
4. Drive key
5. Crankshaft
6. Bearing insert set
7. Main bearing cap
8. Thrust bearing halves (top)
9. Rear main bearing insert (top)
10. Thrust bearing halves (bottom)
11. Rear main bearing insert (bottom)
12. Rear main bearing cap
13. Crankshaft rear seal retainer housing
14. Rear seal
16. Engine rear plate
18. Flywheel
19. Ring gear
20. Oil pressure sender
21. Starter motor

gear onto crankshaft. Be sure that all bearing surfaces are clean and free of nicks or burrs. Install bearing inserts with oil holes in the cylinder block and bearing shells without oil holes in the main bearing caps. Bearing sets for all journals are alike. Lubricate crankshaft and bearings, then set crankshaft in bearings. Lubricate thrust bearing upper halves (without projections) and insert bearing halves between block and crankshaft with oil groove side toward crankshaft shoulders. Use light grease to stick thrust washer lower halves (10) to the sides of rear main bearing cap (12) with tabs aligned with notches in cap. Install bearing insert (11) in cap, lubricate bearing and install cap in original location. Arrow on side of main bearing cap marked "FW" points to flywheel end of engine. Lubricate threads of main bearing cap retaining screws with engine oil and tighten to torque listed in specifications. Check to be sure that crankshaft turns freely before completing assembly.

CRANKSHAFT FRONT OIL SEAL

670 Model

70. The crankshaft front oil seal (23—Fig. 30) is located in timing gear cover and can be renewed after removing crankshaft pulley (2—Fig. 32). Pry seal from timing cover taking care not to damage cover. Seal lip rides on hub of crankshaft pulley. Inspect pulley hub for excessive wear at point of seal contact. Install new seal with lip toward inside using suitable driver. Lubricate seal lip and crankshaft pulley hub, install pulley on crankshaft and tighten pulley retaining screw. The crankshaft pulley of some models is equipped with a pin which must engage a hole in crankshaft gear. Coat threads of center screw (3) with medium strength "Loctite" or equivalent and tighten to 115 N·m (85 ft.-lbs.) torque.

CRANKSHAFT REAR OIL SEAL

670 Model

71. The crankshaft rear oil seal (14—Fig. 32) can be renewed after removing flywheel and seal retainer housing (13). Install new oil seal with lip toward inside. Press seal into housing until flush with outer surface of retainer housing. If crankshaft is grooved at original oil seal contact surface, new seal can be installed 3 mm (1/8 inch) farther into seal housing. Lubricate seal lip, coat sealing surface of retainer housing with RTV sealer and install retainer housing. Tighten screws attaching oil seal retainer housing to rear of cylinder block to 11 N·m (95 in.-lbs.) and screws attaching oil sump to bottom of retainer housing to 9 N·m (78 in.-lbs.) torque.

FLYWHEEL

670 Model

72. To remove flywheel, first separate between the engine and clutch housing using paragraph 50 as a guide. The engine does not need to be separated from the front frame and axle. Unbolt and remove flywheel (18—Fig. 32) from rear of crankshaft. The engine rear plate (16) can be unbolted and removed from rear of engine block after removing flywheel.

On some models, a foam seal may be installed around crankshaft rear seal retainer housing (13). Tighten screws attaching engine rear plate (16) to rear of engine to 49 N·m (36 ft.-lbs.) torque and screws attaching flywheel (18) to crankshaft to 59 N·m (44 ft.-lbs.) torque.

OIL SUMP

670 Model

73. The cast oil sump (15—Fig. 33) is provided with a lower cover (14), which can be removed for access to the oil pick-up (2). If the cast sump must be removed, drain engine oil, then refer to paragraph 50 and remove engine. Unbolt and remove oil sump from engine. Oil pick-up (2) is attached to bottom of cylinder block. Seal (13) is installed at top of oil pump pick-up and retaining screw should be tightened to 11 N·m (96 in.-lbs.) torque.

Clean and dry sealing surface of oil sump and cylinder block, then coat sealing surface with RTV sealer and install oil sump. Tighten screws retaining oil sump to cylinder block to 11 N·m (96 in.-lbs.) torque and screws retaining sump to timing gear housing to 9 N·m (78 in.-lbs.) torque. If removed, tighten screws retaining lower cover (14) to 11 N·m (96 in.-lbs.) torque.

OIL PUMP AND PRESSURE RELIEF VALVE

670 Model

74. The gerotor type oil pump is mounted on front of engine timing gear housing and gear (5—Fig. 29) on pump shaft is driven by crankshaft gear (1).

To remove oil pump, first remove timing gear cover as outlined in paragraph 62, then measure backlash between oil pump gear (5) and crankshaft gear (1). Backlash should be 0.11-0.19 mm (0.0043-0.0075 inch), but should not exceed 0.25 mm (0.010 inch). Unbolt and remove pump.

Pump end cover (7—Fig. 30) can be removed from pump body (11) to inspect pump. End clearance of rotors in pump body should not exceed 0.25 mm (0.10

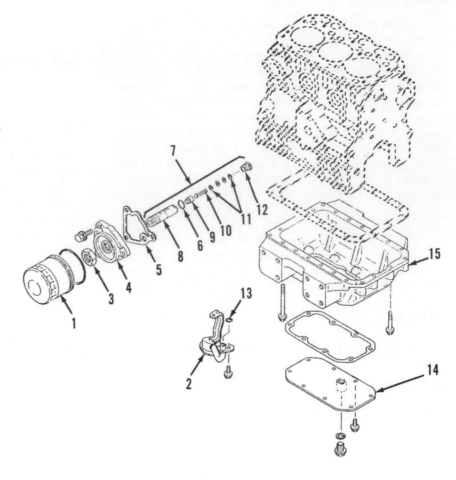

Fig. 33—View of oil sump and relief valve used on 670 model.

1. Oil filter
2. Oil pick-up
3. Nut
4. Filter base
5. Gasket
6. "O" ring
7. Relief valve
8. Adapter body
9. Relief valve poppet
10. Spring
11. Adjusting shims
12. Valve retainer
13. "O" seal
14. Cover
15. Sump

inch). Clearance between outer rotor (10) and pump housing should not exceed 0.25 mm (0.10 inch). Clearance between tip of inner rotor (9) and center of outer rotor lobe should not exceed 0.15 mm (0.006 inch). Pump drive gear (5) can be pressed from shaft (8) if renewal is required, but service pump is available with drive gear. Mark on side of outer rotor (10) should be toward inside of pump housing bore.

Install pump using new gasket and tighten retaining screws to 25 N·m (18 ft.-lbs.) torque. Refer to paragraph 62 for installing timing gear housing.

Oil pressure relief valve (7—Fig. 33) is located inside threaded oil filter adapter. Relief valve pressure can be adjusted by removing valve retainer (12) and adding shims (11). Adding or removing one 1 mm (0.039 inch) thick shim will change oil pressure about 10.9 kPa (1.6 psi).

Free length of spring (10) is 43.5-48.5 mm (1.71-1.91 inch) and spring should exert 20.5 N (4.6 lbs.) when compressed to 27.5 mm (1.08 inch). Relief valve poppet (9) should move freely in bore of body (8) and should not stick. It is not necessary to remove nut (3) to service relief valve, but if nut is loosened, tighten to 30 N·m (22 ft.-lbs.) torque.

ENGINE MODELS
770, 870, 970 AND 1070

75. Yanmar, direct injection, diesel engines are used in 770, 870, 970 and 1070 models. Model 770 tractors are equipped with 3TN82-RJX engine, Model 870 tractors are equipped with 3TN84-RJX engine, Model 970 tractors are equipped with 4TN842-RJX engine, and Model 1070 tractors are equipped with 4TN84-RJX engine. The engines installed in 770 and 870 models have three cylinders; while the engine installed in 970 and 1070 models have four cylinders. The four cylinder engines are equipped with balancer assemblies.

MAINTENANCE

770, 870, 970 And 1070 Models

76. LUBRICATION. Recommended engine lubricant is good quality engine oil with API classification CD/SF or CD/SE. Select oil viscosity depending on ambient temperature. Recommended oil viscosity and ambient temperature ranges are as follows.

Arctic Oil	–55° to 0° C
	(–60° to 32° F)
SAE 5W30	–30° to +10° C
	(–20° to +50° F)
SAE 10W	–20° to +10° C
	(–4° to +50° F)
SAE 10W30	–20° to +20° C
	(–4° to +68° F)
SAE 15W30	–15° to +30° C
	(+5° to +86° F)
SAE 15W40	–15° to +40° C
	(+5° to +104° F)
SAE 20W40	–10° to +40° C
	(+14° to +104° F)
SAE 30	0° to +40° C
	(+32° to +104° F)
SAE 40	+10° to +122° C
	(+50° to +122° F)

Oil and filter should be changed after the first 50 hours of operation for new or rebuilt engine. Thereafter, oil and filter should be changed every 200 hours of normal operation. Operate tractor until oil is warm, then drain engine oil and remove oil filter. Install new filter and refill crankcase with new oil.

Capacity of engine crankcase, including filter, is 4.0 L (4.2 qt.) for 770 model, 4.8 L (5.1 qt.) for 870 model, and 5.3 L (5.6 qt.) for 970 and 1070 models. Oil should be maintained between marks on dipstick of all models.

77. AIR FILTER. The air filter on 770 model should be checked and cleaned, if necessary, after every 200 hours of operation. Remove dust from filter element by tapping lightly with your hand. Low pressure (less that 210 kPa or 30 psi) can be directed through filter from the inside, toward outside, to blow dust from filter. Be careful not to damage filter element while attempting to clean. If special air cleaner solution and water are used to wash air cleaner, make sure element is dry before installing.

NOTE: Do not wash filter with any petroleum solvent.

The two stage air filter on 870, 970 and 1070 models includes a large primary filter element and a smaller secondary element located inside the primary element. The primary filter should be renewed after every 400 hours of operation, when engine looses power, when engine begins to smoke excessively, or at least once every two years. A new secondary filter element should be installed with every third primary element, when dirt can be seen on secondary element, or at least every two years.

A new element should be installed if condition of old element is questionable, regardless of length of service. Filters should be serviced more frequently if operating in extremely dusty conditions.

78. WATER PUMP/FAN/ALTERNATOR BELT. Fan belt tension should be maintained so belt will deflect approximately 13 mm (½ inch) when moderate thumb pressure is applied to belt midway between pump and alternator pulleys. Reposition alternator to adjust belt tension.

79. FUEL FILTER. A renewable paper filter element type fuel filter is located between fuel feed pump and fuel injection pump. Check the clear filter sediment bowl daily for evidence of water or sediment. Bowl should be drained and new filter installed at least every 200 hours of operation or if filter appears dirty. Air must be bled from filter and fuel system as outlined in paragraph 111 each time filter is removed.

R & R ENGINE ASSEMBLY

770 Model

80. To remove engine from tractor, first remove any weights or front mounted equipment. Remove dash lower cover, then unbolt and remove hood. Remove engine side panels, disconnect battery and remove electric starter. Close fuel shut-off at fuel filter, disconnect fuel line at filters, drain fuel from tank, then cover all openings to prevent the entrance of dirt. Disconnect fuel gauge wires and fuel return hose at top of tank. Remove fuel tank mounting screws, mounting brackets and other equipment which would retain tank, then lift tank from tractor. Disconnect throttle control rod. Drain coolant, disconnect or remove radiator hoses, remove muffler and remove air cleaner assembly, then cover all openings. Lower rockshaft lift arms, disconnect or remove hydraulic lines which would interfere with the removal of engine. Disconnect steering drag link (manual steering) or detach power steering hoses. Be sure to always cover hydraulic system openings to prevent the entrance of dirt. Disconnect wire connectors to alternator, oil pressure gauge, temperature gauge and fuel shut-off solenoid. Remove the radiator support rod and coolant recovery tank (located at front of engine). Remove the air shield and the fuel tank support bracket (behind engine). Detach clamp holding wiring harness to top of engine, then move wiring and cables out of the way. Remove the front axle driveshaft from models so equipped.

Remove rocker arm cover from engine, attach two lifting eyes to engine, then attach hoist to lifting eyes. Wedge between front axle and frame to prevent tipping, then stabilize front axle and frame with floor jack under front of frame. Block rear wheels to prevent rolling and support rear of tractor under clutch housing. **Make sure that the rear of tractor, the engine and the front axle, radiator and front frame are all three separately, safely and securely supported.**

Remove screws attaching front frame rails to engine (six screws are located on each side), then carefully roll front frame, radiator and front axle assembly forward away from tractor.

Remove the eight screws attaching engine to clutch housing, shift transmission to neutral, then move the supported engine away from the clutch housing.

Reassemble by reversing removal procedure. Coat splines of clutch shaft lightly with a multipurpose grease before assembling. It may be necessary to turn flywheel to align clutch splines. Tighten the eight screws attaching engine to clutch housing to 54 N·m (40 ft.-lbs.) torque. Apply medium strength "Loctite" or equivalent to threads of the twelve screws attaching front frame to sides of engine before tightening to 90 N·m (65 ft.-lbs.) torque. Adjust hood latches to provide 10 mm (0.394 inch) hood clearance. Adjust by turning screws after loosening latch clamp screws. Refer to paragraph 111 for bleeding the fuel system.

870, 970 And 1070 Models

81. To remove engine from tractor, first remove any weights or front mounted equipment. Remove dash lower cover, then unbolt and remove hood. Remove

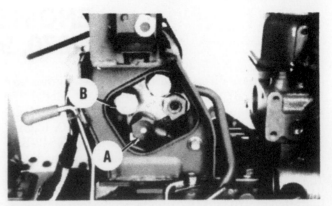

Fig. 34—View of power steering lines attached to steering control valve. Banjo connector (A) incorporates filter.

engine side panels, disconnect battery and remove electric starter. Drain coolant, disconnect or remove radiator hoses, remove muffler and remove air cleaner assembly, then cover all openings. Lower rockshaft lift arms, disconnect or remove hydraulic lines to and from hydraulic pumps, which would interfere with the removal of engine. Disconnect steering drag link (manual steering) or detach power steering lines (A and B—Fig. 34) from steering control valve and move or remove hoses so they will not be damaged when removing engine. Be sure to always cover hydraulic system openings to prevent the entrance of dirt. Disconnect wire connectors to alternator, oil pressure gauge, temperature gauge and fuel shut-off solenoid. Disconnect hourmeter cable and tie cable back, out of the way. Close fuel shut-off at fuel filter, disconnect fuel line at filters, then cover all openings to prevent the entrance of dirt. Disconnect throttle control rod from pump and remove assembly from side of console. Remove the radiator support rod (located at front of engine) and the air shield (behind engine). Detach clamp holding wiring harness at top of engine, then move wiring and cables out of the way. Remove the front axle driveshaft from models so equipped.

Remove rocker arm cover from engine, attach two lifting eyes to engine, then attach hoist to lifting eyes. Wedge between front axle and frame to prevent tipping, then stabilize front axle and frame with floor jack under front of frame. Block rear wheels to prevent rolling and support rear of tractor under clutch housing. **Make sure that the rear of tractor, the engine and the front axle, radiator and front frame are all three separately, safely and securely supported.**

Remove screws attaching front frame rails to engine (six screws are located on each side), then carefully roll front frame, radiator and front axle assembly forward away from tractor.

Remove the eight screws attaching engine to clutch housing, shift transmission to neutral, then move the supported engine away from the clutch housing.

Reassemble by reversing removal procedure. Coat splines of clutch shaft lightly with a multipurpose grease before assembling. It may be necessary to turn flywheel to align clutch splines. Tighten the eight screws attaching engine to clutch housing to 113 N·m (83 ft.-lbs.) torque. Apply medium strength "Loctite" or equivalent to threads of the twelve screws attaching front frame to sides of engine before tightening to 90 N·m (65 ft.-lbs.) torque. Adjust hood latches to provide 10 mm (0.394 inch) hood clearance. Adjust by turning screws after loosening latch clamp screws. Refer to paragraph 111 for bleeding the fuel system.

CYLINDER HEAD

770, 870, 970 And 1070 Models

82. REMOVE AND REINSTALL. To remove cylinder head, drain coolant and engine oil. Remove water pump, rocker arm cover, rocker arm assembly,

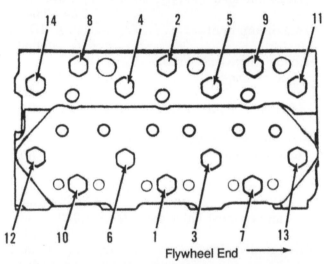

Fig. 35—Cylinder head retaining screws for three cylinder engines should be tightened in at least three stages in the sequence shown. Refer to text for recommended torque values for each stage.

valve caps and push rods. Remove fuel injection lines and fuel return lines. The inlet manifold, exhaust manifold, glow plugs and injection nozzles do not need to be removed to remove cylinder head, but if required for other service, these parts should be removed before removing head. Remove retaining screws in reverse of the order shown in Fig. 35 or Fig. 36, then lift cylinder head from engine.

Refer to appropriate paragraphs for servicing valves, guides and springs. Clean cylinder block and cylinder head gasket surfaces and inspect cylinder head for cracks, distortion or other defects. If alignment pins were removed with cylinder head, remove pins and install in cylinder block. Check head gasket surface with a straight edge and feeler gauge. Recondition or renew cylinder head if distortion exceeds 0.15 mm (0.006 inch). Cylinder head thickness should not be reduced more than 0.20 mm (0.008 inch).

Place cylinder head gasket over alignment pins and make sure that oil passage in gasket is positioned over passage in block. Lubricate bolts with engine oil and install bolts, tightening in three steps in the sequence shown in Fig. 35 or Fig. 36. Refer to the following for correct three step torques.

First torque . . . . 24.2 N·m
(18.0 ft.-lbs.)
Second torque. . . . . 48.4 N·m
(36.0 ft.-lbs.)
Third (final) torque . . . . 78 N·m
(58 ft.-lbs.)

Tighten stud nuts retaining rocker arm assembly to 26 N·m (19 ft.-lbs.) torque. Tighten rocker arm cover retaining nuts to 18 N·m (160 in.-lbs.) torque. Tighten nozzle retaining plate nuts to 4.5 N·m (39 in.-lbs.) torque and nozzle leak-off line nuts to 15 N·m (130 in.-lbs.) torque. Tighten screws retaining inlet manifold and exhaust manifold to 26 N·m (19 ft.-lbs.) torque.

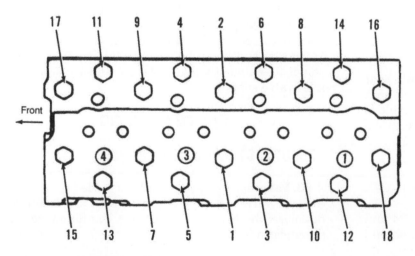

Fig. 36—Cylinder head retaining screws of four cylinder models should be tightened in at least three stages in the sequence shown. Refer to text for recommended torque values for each stage.

VALVE CLEARANCE

770 And 870 Models

83. Valve clearance should be adjusted with engine cold. Specified clearance is 0.2 mm (0.008 inch) for both inlet and exhaust valves. Adjustment is make by loosening locknut and turning adjusting screw (18—Fig. 37) at push rod end of rocker arm.

To adjust valve clearance, turn crankshaft in normal direction of rotation so that rear (number 1) piston is at top dead center of compression stroke. Adjust clearance of inlet and exhaust valves for rear cylinder. On three cylinder engines, turn crankshaft 240° until next piston in firing order (front) is at top dead center and adjust clearance for both valves of that cylinder. Turn crankshaft another 240° until last piston in firing order (center) is at top dead center, then adjust valve clearance for both valves of the last cylinder.

970 And 1070 Models

84. Valve clearance should be adjusted with engine cold. Specified clearance is 0.2 mm (0.008 inch) for both inlet and exhaust valves. Adjustment is make by loosening locknut and turning adjusting screw (18—Fig. 37) at push rod end of rocker arm.

To adjust valve clearance, turn crankshaft in normal direction of rotation so that rear (number 1) piston is at top dead center of compression stroke. Adjust clearance of inlet and exhaust valves for rear cylinder. Cylinders are numbered from rear to front and firing order is 1-3-4-2. Turn crankshaft 180° until next piston in firing order (3) is at top dead center and adjust clearance for both valves of that cylinder. Turn crankshaft another 180° until next piston in firing order (4-front) is at top dead center and adjust valve clearance for that cylinder. Turn crankshaft 180° until the last piston in firing order (2) is at top dead center, then adjust valve clearance for both valves of the last cylinder.

VALVES, GUIDES AND SEATS

770, 870, 970 And 1070 Models

85. Both inlet and exhaust valves originally seat directly in cylinder head. Inlet valve face and seat angle is 30° and exhaust valve face and seat angle is 45°. Refer to the following recommended valve, guide and seat specifications.

770
Engine Model 3TN82-RJX
Inlet valve seat width —
   Recommended . . . . 1.07-1.24 mm
                                                            (0.042-0.049 in.)

Maximum limit . . . . 1.74 mm
                                                            (0.069 in.)
Exhaust valve seat width —
   Recommended . . . . 1.24-1.45 mm
                                                            (0.049-0.057 in.)
   Maximum limit . . . . 1.94 mm
                                                            (0.076 in.)
Inlet and exhaust valve seat recession —
   Recommended . . . . 0.30-0.50 mm
                                                            (0.012-0.020 in.)
Inlet and exhaust valve stem OD . . 7.960-7.975 mm
                                                            (0.3134-0.3140 in.)
Inlet and exhaust valve guide ID . . 8.010-8.030 mm
                                                            (0.3154-0.3160 in.)
Valve guide height. . . . . 15.00 mm
                                                            (0.591 in.)

870, 970 and 1070
Engine Models 3TN84-RJX, 4TN842-RJX
                                             and 4TN84-RJX
Inlet valve seat width —
   Recommended . . . . 1.07-1.24 mm
                                                            (0.042-0.049 in.)
   Maximum limit . . . . 1.74 mm
                                                            (0.069 in.)
Exhaust valve seat width —
   Recommended . . . . 1.24-1.35 mm
                                                            (0.049-0.053 in.)
   Maximum limit . . . . 1.94 mm
                                                            (0.076 in.)
Inlet and exhaust valve seat recession —
   Recommended . . . . 0.30-0.50 mm
                                                            (0.012-0.020 in.)
Inlet and exhaust valve stem OD . . 7.960-7.975 mm
                                                            (0.3134-0.3140 in.)
Inlet and exhaust valve guide ID . . 8.010-8.030 mm
                                                            (0.3154-0.3160 in.)
Valve guide height. . . . . 15.00 mm
                                                            (0.591 in.)

Valve stem to guide clearance should not exceed 0.15 mm (0.006 inch); however, guides can be knurled to reduce clearance if less than 0.20 mm (0.008 inch). Install new guide if clearance exceeds 0.20 mm (0.008 inch) with new valve. Press new guide into cylinder head until height of guide above surface of cylinder head is correct as listed in preceding specifications. Valve seat width can be narrowed using 60° and 30° stones for exhaust valve seats; 45° and 15° stones for inlet valve seats.

VALVE SPRINGS

770, 870, 970 And 1070 Models

86. Valve springs are interchangeable for inlet and exhaust valves. Renew springs which are distorted, discolored by heat or fail to meet the following test specifications.

770
Engine Model. 3TN82-RJX
Valve spring free length . . . . 40 mm
(1.575 in.)

Pressure at compressed
height . . . . 319 N @ 24.0 mm
(72 lbs. @ 0.95 in.)

870, 970 and 1070
Engine Models 3TN84RJX, 4TN842-RJX
and 4TN84-RJX
Valve spring free length . . . . 41.5 mm
(1.634 in.)

Pressure at compressed
height . . . . 319 N @ 24 mm
(72 lbs. @ 0.945 in.)

ROCKER ARMS AND PUSH RODS

770, 870, 970 And 1070 Models

87. The rocker arms assembly (Fig. 37) can be unbolted and removed after removing the rocker arm cover. Push rods can be withdrawn after removing rocker arms. Inlet and exhaust rocker arms are alike, but all valve train components should be identified as

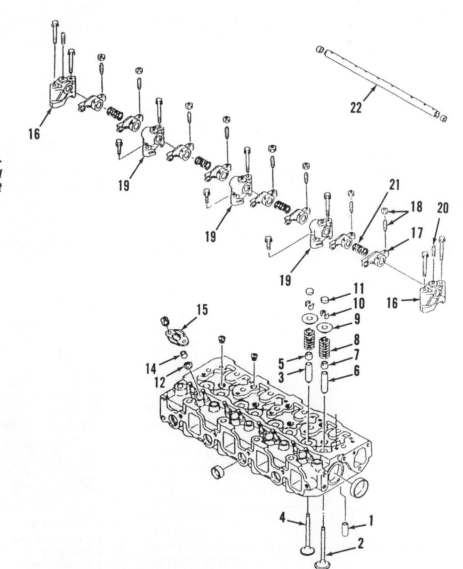

Fig. 37—Exploded view of typical cylinder head. Valve stem seal (7) and guide (6) for inlet valves are different from similar parts for exhaust valves.

1. Alignment dowel
2. Inlet valve
3. Exhaust valve guide
4. Exhaust valve
5. Stem seal (exhaust)
6. Inlet valve guide
7. Stem seal (inlet)
8. Valve spring
9. Spring seat
10. Split retainer
11. Valve cap
12. Seal washer
14. Seal
15. Nozzle retainer
16. End support
17. Rocker arm
18. Adjuster screw and locknut
19. Center support
20. Set screw
21. Spacer spring
22. Shaft

they are removed so they can be reinstalled in original locations if reused. Install new component if old part does not meet standard listed in the following specifications.

770
Engine Model	3TN82-RJX

Rocker shaft OD —
Desired .... 15.966-15.984 mm
(0.6286-0.6293 in.)
Wear limit .... 15.96 mm
(0.628 in.)

Shaft support ID —
Desired .... 16.0-16.02 mm
(0.630-0.631 in.)
Wear limit .... 16.09 mm
(0.633 inch)

Rocker arm ID —
Desired .... 16.0-16.02 mm
(0.630-0.631 in.)
Wear limit .... 16.09 mm
(0.633 in.)

Rocker arm to shaft clearance —
Desired .... 0.016-0.054 mm
(0.0006-0.0021 in.)
Wear limit .... 0.13 mm
(0.005 in.)
Push rod runout, Max. .... 0.075 mm
(0.003 in.)
Push rod length, Min. .... 178.25-178.75 mm
(7.018-7.037 in.)

870, 970 and 1070
Engine Models 3TN84-RJX, 4TN842-RJX
and 4TN84-RJX

Rocker shaft OD —
Desired .... 15.966-15.984 mm
(0.6286-0.6293 in.)
Wear limit .... 15.96 mm
(0.628 in.)

Shaft support ID —
Desired .... 16.0-16.02 mm
(0.630-0.631 in.)
Wear limit .... 16.09 mm
(0.633 in.)

Rocker arm ID —
Desired .... 16.0-16.02 mm
(0.630-0.631 in.)
Wear limit .... 16.09 mm
(0.633 in.)

Rocker arm to shaft clearance —
Desired .... 0.016-0.054 mm
(0.0006-0.0021 in.)
Wear limit .... 0.13 mm
(0.005 in.)
Push rod runout, Max. .... 0.075 mm
(0.003 in.)
Push rod length, Min. .... 178.25-178.75 mm
(7.018-7.037 in.)

Assemble rocker arms, springs and the two center supports. Align set screw (20—Fig. 37) in the two supports at ends with holes in shaft, then tighten set screws. Tighten stud nuts retaining rocker arm assembly to 26 N•m (226 in.-lbs.) torque and adjust valve clearance as outlined in paragraph 57. Tighten rocker arm cover retaining nuts to 18 N•m (160 in.-lbs.) torque.

CAM FOLLOWERS

770, 870, 970 And 1070 Models

88. Cam followers can be removed after removing rocker arms assembly, push rods (as outlined in paragraph 87) and the engine camshaft (as outlined in paragraph 95). Invert engine to hold cam followers away from camshaft before removing camshaft. Remove camshaft, then lift cam followers from block bores. Cam followers must be identified as they are removed so that they can be reinstalled in same location. Minimum outside diameter of cam follower is 11.93 mm (0.470 inch). Maximum inside diameter of cam follower bore in cylinder block is 12.05 mm (0.474 inch). Cam follower to bore clearance should be 0.010-0.043 mm (0.0003-0.0016 inch), but should not exceed 0.10 mm (0.004 inch).

Fig. 38—View of timing gears typical of 770, 870, 970 and 1070 models.

1. Crankshaft gear
2. Idler gear
3. Camshaft gear
4. Fuel injection pump gear
5. Oil pump drive gear

TIMING GEARS

770, 870, 970 And 1070 Models

89. The timing gear train consists of crankshaft gear (1—Fig. 38), idler gear (2), camshaft gear (3) and fuel injection pump drive gear (4). Timing gear marks (T1, T2 and T3) are visible after removing the timing gear cover as outlined in paragraph 93. Valves and injection pump are properly timed to crankshaft when marks on gears are aligned as shown in Fig. 38. Because of the number of teeth on idler gear, timing marks will not align every other revolution of the crankshaft. Alignment can be checked by turning crankshaft until keyway is vertical, then checking location of mark on camshaft gear. If mark on camshaft gear is away from idler gear, turn crankshaft one complete turn until keyway is again vertical. Remove idler gear, then reinstall idler gear with marks on all gears aligned as shown in Fig. 38.

NOTE: Before removing any gears, first refer to paragraph 87 and remove rocker arm assembly to avoid possible damage to piston or valve train. Damage could result if either camshaft or crank-

shaft is turned independently from the other unless rocker arms are removed.

Backlash between timing gears (1, 2, 3 or 4) and any one of the other timing gears should be less than 0.20 mm (0.008 inch). Backlash between oil pump gear (5) and crankshaft gear (1) should also be less than 0.20 mm (0.008 inch).

All timing gears (1, 2, 3 and 4) have a single timing mark at (T1, T2 and T3). Marks on idler gear (2) are located on teeth and marks on timing gears (1, 3 and 4) are located in tooth valley.

90. IDLER GEAR AND SHAFT. To inspect or remove the idler gear (2—Fig. 39) and shaft (12), first remove timing gear cover as outlined in paragraph 93. Check backlash between idler gear and other timing gears before removing the gear. Backlash between idler gear and any other gear (1, 3 or 4) should be less than 0.20 mm (0.008 inch).

NOTE: Before removing any gears, first refer to paragraph 87 and remove rocker arm assembly to avoid possible damage to piston or valve train. Damage could result if either camshaft or crank-

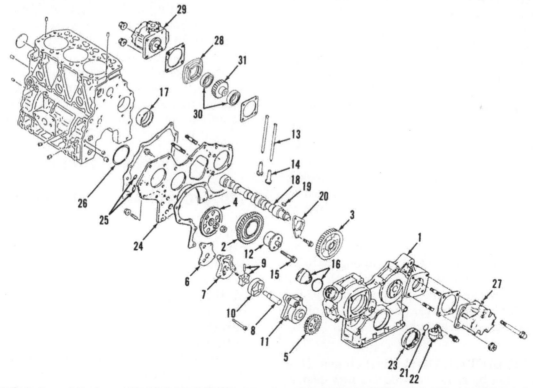

Fig. 39—Exploded view of timing gear cover and timing gear housing typical of 770, 870, 970 and 1070 models.

1. Timing gear cover	9. Inner rotor & drive pin	17. Camshaft bearing	24. Timing gear housing
2. Idler gear	10. Outer rotor	18. Valve camshaft	25. "O" rings
3. Camshaft drive gear	11. Pump housing	19. Drive key	26. "O" ring
4. Injection pump drive gear	12. Idler shaft	20. Thrust plate	27. Power steering pump
5. Oil pump drive gear	13. Push rod	21. "O" ring	28. Hydraulic pump adapter
6. Gasket	14. Cam follower	22. Tachometer gear unit	29. Hydraulic pump
7. Oil pump cover	15. Retaining screws	23. Crankshaft front seal	30. Bearings
8. Pump shaft	16. Filler cap and gasket		31. Drive gear

shaft is turned independently from the other unless rocker arms are removed.

Remove the two screws (15) retaining idler shaft to front of cylinder block, then withdraw gear (2) and idler shaft (12). Inside diameter of idler gear bushing should be 46.0-46.025 mm (1.811-1.812 inch), but should not exceed 46.08 mm (1.814 inch). Outside diameter of shaft should be 45.950-45.975 mm (1.809-1.810 inch), but should not be less than 45.93 mm (1.808 inch). Renew shaft and/or bushing inside gear if clearance exceeds 0.15 mm (0.006 inch). Oil hole in bushing must be aligned with hole in idler gear. Inspect gear teeth for wear or scoring.

To install idler gear, position crankshaft with rear (number 1) piston at top dead center. Turn camshaft gear (3—Fig. 38) and fuel injection pump drive gear (4) so that timing marks point to center of idler gear shaft. Install idler gear (2) and idler shaft so that all timing marks are aligned as shown in Fig. 38, then install idler shaft retaining screws. Tighten retaining screws to 26 N·m (230 in.-lbs.) torque.

91. CAMSHAFT GEAR. The camshaft must be removed as outlined in paragraph 95 before gear (3—Fig. 39) can be pressed from shaft. Check camshaft end play and backlash between camshaft gear (3) and idler gear (2) before removing camshaft. Backlash should be 0.04-0.12 mm (0.0016-0.0047 inch), but should not exceed 0.20 mm (0.008 inch). Camshaft end play should be 0.05-0.20 mm (0.002-0.008 inch), but should not exceed 0.4 mm (0.016 inch). If end play is excessive, install new thrust plate (20).

Remove rocker arm assembly as outlined in paragraph 87 and idler gear as outlined in paragraph 90. Damage could result if either camshaft or crankshaft is turned independently from the other unless rocker arms are removed. Remove camshaft as outlined in paragraph 95. Gear can be pressed from camshaft to renew thrust plate (20) or gear (3).

When reinstalling camshaft in all models, tighten screws retaining thrust plate to front of engine to 26 N·m (230 in.-lbs.) torque. Refer to paragraph 87 for installation of rocker arm assembly, paragraph 93 for installation of timing gear cover and to paragraph 83 or 84 for adjusting valve clearance.

92. CRANKSHAFT GEAR. Crankshaft gear (1—Fig. 39) should not be removed unless a new gear is to be installed. Backlash between crankshaft gear (1) and idler gear (2) should be 0.11-0.19 mm (0.0043-0.0075 inch), but should not exceed 0.20 mm (0.008 inch). Backlash between crankshaft gear (1) and oil pump gear (5) should be within the same limits.

To remove crankshaft gear (1), the crankshaft should be removed and gear pressed from shaft.

92. INJECTION PUMP DRIVE GEAR. Refer to paragraph 116 for removal or other service procedures to the injection pump drive gear (4—Fig. 39).

TIMING GEAR COVER

770, 870, 970 And 1070 Models

93. Crankshaft front oil seal (23—Fig. 39) can be removed and new seal installed after removing crankshaft pulley. Seal should be driven into cover until flush with front of cover. Lubricate seal before installing crankshaft pulley.

To remove timing gear cover (1), first refer to paragraph 80 or 81 and separate front end of tractor from the engine. Remove fan and alternator. Remove retaining screw from center of crankshaft pulley. Notice that screw in center of crankshaft pulley should have "Loctite" or equivalent on threads and may be difficult to loosen. Use an appropriate puller and remove crankshaft pulley from crankshaft. Unbolt and remove timing gear cover from front of engine.

Clean mating surfaces of timing gear cover and timing gear housing, then apply an even bead of sealer to mating surface. A gasket is not used between cover and housing. Install cover and tighten retaining screws to 26 N·m (230 in.-lbs.) torque. The crankshaft pulley of some models is equipped with a pin which must engage a hole in crankshaft gear. Coat threads of center screw with medium strength "Loctite" or equivalent and tighten to 115 N·m (85 ft.-lbs.) torque. Remainder of assembly is reverse of disassembly.

TIMING GEAR HOUSING

770, 870, 970 And 1070 Models

94. To remove timing gear housing (24—Fig. 39), first separate front end of tractor from the engine as outlined in paragraph 80 or 81, the rocker arm cover and rocker arm assembly as outlined in paragraph 87, timing gear cover (1) as outlined in paragraph 93, camshaft with drive gear (3 and 18) as outlined in paragraph 95, and injection pump drive gear (4) as outlined in paragraph 116. Unbolt and remove oil pump (5 through 11) from timing gear housing and oil pan (sump) from bottom of engine, then unbolt timing gear housing (24) from front of cylinder block.

When reinstalling timing gear housing, reverse removal procedure. Clean and dry sealing surfaces of timing gear housing and cylinder block, then apply bead of RTV sealer on block mating surface of timing gear housing. Install timing gear housing and tighten retaining screws to 20 N·m (177 in.-lbs.) torque.

CAMSHAFT

770, 870, 970 And 1070 Models

95. To remove the engine camshaft, first remove rocker arm cover and rocker arm assembly as outlined in paragraph 87 and timing gear cover as outlined in paragraph 93. Check camshaft end play and backlash between camshaft gear (3—Fig. 39) and idler gear (2) before removing camshaft. Backlash should be 0.04-0.12 mm (0.0016-0.0047 inch), but should not exceed 0.20 mm (0.008 inch). Camshaft end play should be 0.05-0.20 mm (0.002-0.008 inch), but should not exceed 0.4 mm (0.16 inch). If end play is excessive, install new thrust plate (20). Remove and invert engine to hold cam followers (14) away from camshaft (18).

NOTE: It may not be necessary to turn engine up-side-down if magnets are available to hold all of the cam followers away from camshaft.

Remove camshaft thrust plate retaining screws through the web holes of camshaft gear (3), then pull camshaft (18) out of engine block bores. Gear can be pressed from camshaft to renew thrust plate (20) or gear. Flywheel must be removed to remove plug from rear of camshaft bore in cylinder block. Refer to the following specifications.

770, 870, 970 and 1070 Models
Camshaft journal OD —

Front	44.925-44.950 mm
	(1.769-1.770 in.)
Wear limit	44.80 mm
	(1.764 in.)
Intermediate	44.910-44.935 mm
	(1.768-1.769 in.)
Wear limit	44.80 mm
	(1.764 in.)
Rear	44.925-44.950 mm
	(1.769-1.770 in.)
Wear limit	44.80 mm
	(1.764 in.)
Camshaft bearings, ID	44.990-45.055 mm
	(1.771-1.774 in.)
Wear limit	45.10 mm
	(1.776 in.)
Camshaft end play	0.05-0.20 mm
	(0.002-0.008 in.)
Wear limit	0.4 mm
	(0.16 in.)
Cam lobe height	38.635-38.765 mm
	(1.521-1.526 in.)
Wear limit	38.40 mm
	(1.512 in.)

Use 1¾-inch piloted driver to install new bushing (17), and be sure to align oil hole in bushing with

passage in cylinder block. Cam gear (3) should be heated to approximately 150° C (300° F) before pressing onto camshaft. Be sure that thrust plate (20) is installed and will turn freely after gear is pressed firmly against shoulder of camshaft.

When reinstalling camshaft in all models, tighten screws retaining thrust plate to front of engine to 26 N·m (230 in.-lbs.) torque. Refer to paragraph 87 for installation of rocker arm assembly, paragraph 93 for installation of timing gear cover and to paragraph 83 or 84 for adjusting valve clearance.

ROD AND PISTON UNITS

770, 870, 970 And 1070 Models

96. Piston and connecting rod units are removed from above after removing oil pan and cylinder head.

Drain engine oil and coolant, then refer to paragraph 80 or 81 and remove the engine assembly. Remove the cylinder head as outlined in paragraph 82. Remove the oil sump cover and, on four cylinder models, refer to paragraph 107 and remove the balancer assembly. On all models, remove the ridge (if present) from top of cylinder bore. Check to make sure piston, rod and cap are marked before removing pistons. If not already marked, stamp cylinder number on piston, rod and cap before removing. Number "1" cylinder is at rear and connecting rods should be marked on injection pump (right) side. Keep rod bearing cap with matching connecting rod.

Piston must be assembled to connecting rod with recess in top of piston opposite slot in connecting rod for bearing insert tang. Screws attaching connecting rod cap to rod should not be reused. Always install **new** connecting rod screws.

Lubricate cylinder, crankpin and ring compressor with clean engine oil. Space piston ring end gaps evenly around piston, but not aligned with piston pin, then compress piston rings with suitable compressor tool. Install piston and rod units in cylinder block with recess in top of piston on same side as fuel injection pump (right side). Make sure that bore in connecting rod and cap is completely clean. Install bearing inserts with tangs engaging slot in rod and cap. Make sure that inserts are firmly seated, lubricate crankpin and bearing inserts, then seat connecting rod with insert against crankpin. Install connecting rod cap with insert firmly seated and lubricated. Tangs on both bearing inserts should be on left (camshaft) side of engine. Tighten connecting rod cap retaining screws to 47 N·m (35 ft.-lbs.) torque.

Clearance between flat top of piston and cylinder head can be measured using 10 mm (0.4 inch) long sections of 1.5 mm (0.06 inch) diameter soft lead wire positioned on top of piston. Turn crankshaft one complete revolution, then remove lead wire and measure thickness of the flattened wire. Clearance should be 0.64-0.82 mm (0.025-0.032 inch).

On four cylinder models, refer to paragraph 107 and install the balancer assembly. On all models, clean mating surfaces of crankcase extension and oil sump cover thoroughly. Apply a bead of "Form-A-Gasket" or equivalent to sealing surfaces and tighten oil sump cover retaining screws to 27 N·m (20 ft.-lbs.) torque. Remainder of assembly is reverse of disassembly.

PISTON, RINGS AND CYLINDER

770, 870, 970 And 1070 Models

97. Pistons are fitted with two compression rings and one oil control ring. Piston pin is full floating and retained by a snap ring at each end of pin bore in piston. If necessary to separate piston from connecting rod, remove snap rings and push pin from piston and connecting rod. Service pistons are available in standard size and 0.25 mm (0.010 inch) oversize.

Inspect all piston skirts for scoring, cracks and excessive wear. On 770 model, measure piston skirt at right angles to pin bore 22 mm (0.866 inch) from bottom of skirt. Minimum skirt measurement for standard size piston is 81.80 mm (3.220 inches).

On 870, 970 and 1070 models, measure piston skirt at right angles to pin bore 29.0 mm (1.142 inch) from bottom of skirt. Minimum skirt measurement for standard size piston is 83.80 mm (3.299 inches).

Piston pin bore in piston should not exceed 26.02 mm (1.024 inch). Clearance between piston pin and bore in piston should not exceed 0.05 mm (0.002 inch).

Standard cylinder bore diameter for 770 model is 82.00-82.03 mm (3.228-3.230 inches) and should not exceed 82.20 mm (3.236 inches). Piston to cylinder bore clearance should not exceed 0.30 mm (0.012 inch). Cylinders can be rebored and fitted with 0.25 mm (0.10 inch) oversize pistons.

Standard cylinder bore diameter for 870, 970 and 1070 models is 84.00-84.03 mm (3.307-3.308 inches) and should not exceed 84.20 mm (3.315 inches). Piston to cylinder bore clearance should not exceed 0.30 mm (0.012 inch). Cylinders can be rebored and fitted with 0.25 mm (0.10 inch) oversize pistons.

Cylinder bore of all models should be deglazed using 180 grit stone and crosshatch pattern should intersect at 30-40°. Wash abrasive from cylinder walls using warm soapy water. Continue cleaning until white cleaning cloths show no discoloration. Make sure all soap is rinsed from cylinders, dry cylinder block thoroughly, then lubricate with clean engine oil.

NOTE: Use only mild soap to clean abrasive from cylinders. Do not use commercial solvents, gasoline or kerosene to clean cylinders.

End gap of piston rings in cylinder should not exceed 1.50 mm (0.059 inch). Rings should be inserted squarely in cylinder using piston to push ring approximately 30 mm (1.181 inch) into cylinder.

Piston ring side clearance in grooves should not exceed 0.25 mm (0.010 inch) for top and second (compression) rings or 0.20 mm (0.008 inch) for oil control (bottom) ring.

When installing piston rings, install oil ring expander in groove, followed by the oil ring. End gap of oil ring should be 180° from ends of expander. Install compression rings with inside chamfer or identification mark toward top of piston. Install chrome compression ring in top groove. Space end gaps of top, second and oil control rings 120° apart before installing in cylinder bore.

PISTON PIN

770, 870, 970 And 1070 Models

98. Piston pin bore in piston should not exceed 26.02 mm (1.024 inch). Clearance between piston pin and bore in piston should not exceed 0.05 mm (0.002 inch).

Piston pin diameter should be measured at each end and in center. Measure pin diameter at each of the three locations and again 90° from each of the original locations. Piston pin diameter should not be less than 25.90 mm (1.020 inch).

Inside diameter of pin bushing in top of connecting rod should not exceed 26.10 mm (1.028 inch). Clearance between piston pin and bore of pin bushing in connecting rod should not exceed 0.11 mm (0.004 inch).

CONNECTING ROD AND BEARINGS

770, 870, 970 And 1070 Models

99. Connecting rod can be unbolted from crankshaft crankpin and new bearing inserts installed from below after removing oil sump cover and balancer, if so equipped. Refer to paragraph 96 for removing connecting rods. Connecting rod should not be out-of-parallel or twisted more than 0.08 mm (0.003 inch).

Inside diameter of pin bushing in top of connecting rod should be 26.025-26.038 mm (1.0246-1.0251 inch), but should not exceed 26.10 mm (1.028 inch). Clearance between piston pin and bushing bore in connecting rod should be 0.025-0.047 mm (0.0010-0.0019 inch), but should not exceed 0.11 mm (0.004 inch).

Connecting rod bearing clearance on crankshaft crankpin should be 0.038-0.090 mm (0.0015-0.0035 inch) and wear limit should not exceed 0.16 mm (0.006 inch). Connecting rod side play on crankpin should be 0.2-0.4 mm (0.0079-0.0157 inch), but should not exceed 0.55 mm (0.022 inch).

Crankpin standard diameter is 47.952-47.962 mm (1.8879-1.8883 inches). Undersized bearing inserts, which require crankshaft journals to be resized and polished, are available for service. Finished undersize journals should provide standard clearance with undersized bearings.

Refer to paragraph 96 if connecting rod and piston units are removed from engine. Make sure that bore in connecting rod and cap is completely clean and install bearing inserts with tangs engaging slot in rod and cap. Make sure that inserts are firmly seated, lubricate crankpin and bearing inserts, then seat connecting rod with insert against crankpin. Install connecting rod cap with insert firmly seated and lubricated. Tangs on both halves of bearing should be on left (camshaft) side of engine. Tighten connecting rod cap retaining screws to 47 N·m (35 ft.-lbs.) torque. If so equipped, refer to paragraph 107 for installing and timing the balancer. Clean and dry sealing surface of block, crankcase extension/oil sump and oil sump cover, then apply a bead of "Form-A-Gasket" or equivalent to sealing surfaces. Install oil sump and tighten retaining screws to 27 N·m (20 ft.-lbs.) torque. Remainder of assembly is reverse of disassembly.

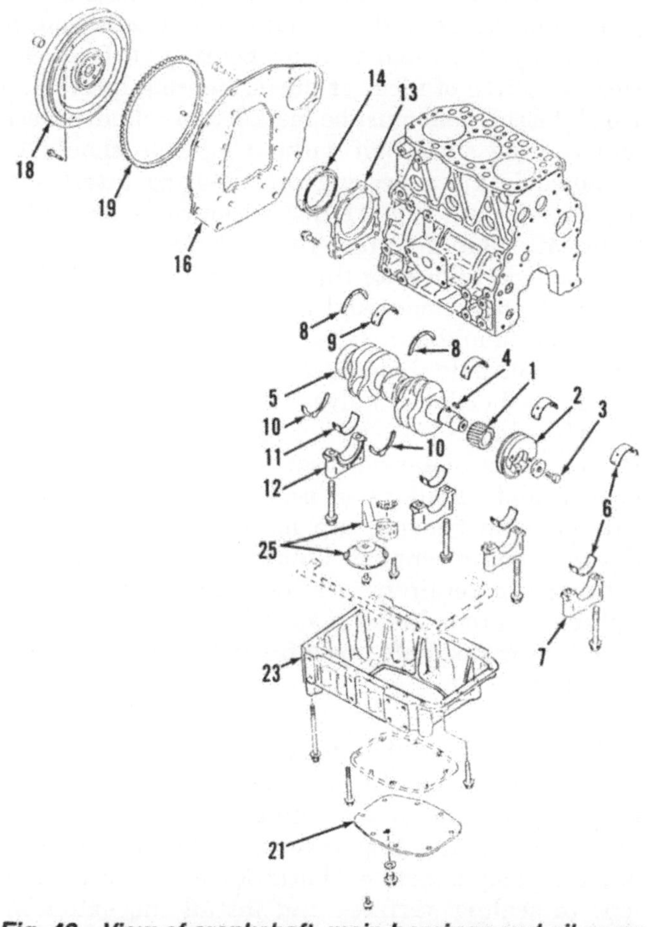

CRANKSHAFT AND MAIN BEARINGS

770, 870, 970 And 1070 Models

100. The crankshaft is supported in four main bearings. Crankshaft end play is controlled by insert type thrust bearing (8 and 10—Fig. 40 or Fig. 41) on sides of rear (number 1) main bearing journal.

To remove crankshaft (5), remove engine from tractor as outlined in paragraph 80 or 81, the rocker arm cover and rocker arm assembly as outlined in paragraph 87, the timing gear cover as outlined in paragraph 93, camshaft as outlined in paragraph 95, injection pump drive gear and camshaft as outlined in paragraph 116 and timing gear housing as outlined in paragraph 94. Remove flywheel (18), then remove crankshaft rear oil seal retainer (13). The three lower screws attaching the oil seal retainer housing (13—Fig. 40 or Fig. 41) and the two screws directly below seal retainer housing are threaded into the crankcase extension and sump (23). Unbolt and remove the sump cover (21), then unbolt and remove the balancer assembly from four cylinder models. On all models, remove screws attaching the crankcase extension and sump to the cylinder block (crankcase) and remove the extension housing.

Unbolt and remove the engine rear plate (16—Fig. 40 or Fig. 41) from all models. Invert engine, then refer to paragraph 96 and detach connecting rods from crankshaft. Mark all main bearing caps for correct installation, unbolt and remove crankshaft main bearing caps, then lift crankshaft from engine. The rear main bearing is number "1."

Check all main bearing journals, thrust surfaces of rear journal and connecting rod crankpin journals for scoring or excessive wear. Main bearings and rod bearings are available in standard size and 0.25 mm (0.010 inch) undersize. Refer to the following specifications.

Fig. 40—View of crankshaft, main bearings and oil sump typical of three cylinder 770 and 870 models. Refer to Fig. 41 for 970 and 1070 models.

1. Crankshaft gear
2. Crankshaft pulley
3. Retaining screw
4. Drive key
5. Crankshaft
6. Bearing insert set
7. Main bearing cap
8. Thrust bearing halves (top)
9. Rear main bearing insert (top)
10. Thrust bearing halves (bottom)
11. Rear main bearing insert (bottom)
12. Rear main bearing cap
13. Crankshaft rear seal retainer housing
14. Rear seal
16. Engine rear plate
18. Flywheel
19. Ring gear
21. Oil sump cover
23. Crankcase extension & sump
25. Oil pick-up

770, 870, 970 and 1070 Models

Crankpin standard OD —
Desired . . . . 47.952-47.962 mm
(1.8879-1.8883 in.)
Wear limit . . . . 47.91 mm
(1.886 in.)

Crankpin to rod bearing clearance —
Desired . . . . 0.038-0.090 mm
(0.0015-0.0035 in.)
Wear limit . . . . 0.16 mm
(0.006 in.)

Main journal standard OD —
Desired . . . . 49.952-49.962 mm
(1.9666-1.9670 in.)
Wear limit . . . . 49.90 mm
(1.965 in.)

Main journal to bearing clearance —
Desired . . . . 0.038-0.093 mm
(0.0015-0.0037 in.)
Wear limit . . . . 0.15 mm
(0.006 in.)

Crankshaft end play —
Desired . . . . 0.090-0.271 mm
(0.004-0.011 in.)
Wear limit. . . . . 0.33 mm
(0.013 in.)
Main cap torque. . . . . 98 N·m
(72 ft.-lbs.)

Rear oil seal case torque —
Seal to block . . . . 26 N·m
(266 in.-lbs.)
Oil pan to seal case . . . . 9 N·m
(78 in.-lbs.)

If crankshaft timing gear (1—Fig. 40 or Fig. 41) or balancer drive gear (20—Fig. 41) was removed, heat gear to approximately 150° C (300° F) before pressing gear onto crankshaft. Be sure that crankshaft is completely clean and that all bearing surfaces are clean and free of nicks or burrs. Bearing insert half which has oil hole must be installed in cylinder block and bearing shell half without oil hole should be installed in main bearing cap. Bearing insert sets (6—Fig. 40 and Fig. 41) for all journals are alike. Lubricate main bearings, then set clean crankshaft in bearings. Lubricate thrust bearing upper halves (without projections) and insert bearing halves between block and crankshaft with oil groove side toward crankshaft shoulders. Use light grease to stick thrust washer lower halves (10) to the sides of rear main bearing cap (12) with tabs aligned with notches in cap. Install bearing insert (11) in cap, lubricate bearing and install cap in original location. Arrow on side of main bearing cap marked "FW" points to flywheel end of engine. Lubricate threads of main bearing cap retaining screws with engine oil and tighten to torque listed in specifications. Check to be sure that crankshaft turns freely before completing assembly.

When installing the crankcase extension (23), observe the following. Clean mating surfaces of extension, engine rear plate, timing gear housing (front plate), the timing gear cover and the oil pan thoroughly. Apply a bead of "Form-A-Gasket" or equivalent to sealing surfaces and install the extension housing. Tighten screws attaching the crankcase extension housing to the cylinder block (crankcase) to 27 N·m (20 ft.-lbs.) torque. Tighten screws attaching timing gear housing to crankcase extension to 22 N·m (16 ft.-lbs.) torque. Tighten screws (B—Fig. 42) to 49 N·m (36 ft.-lbs.) torque and screws (A) to 27 N·m (20 ft.-lbs.) torque. On four cylinder models, refer to paragraph 107 and install balancer assembly with timing marks aligned. Tighten screws retaining oil sump cover (21—Fig. 40 or Fig. 41) to crankcase extension to 27 N·m (20 ft.-lbs.) torque. Remainder of assembly is reverse of disassembly.

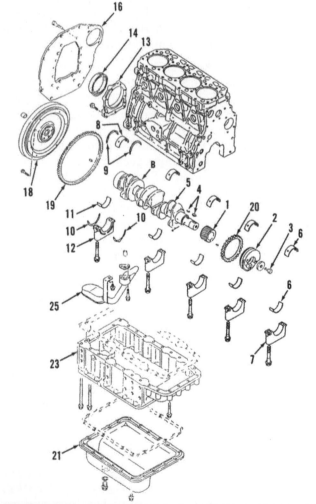

Fig. 41—View of crankshaft, main bearings and oil sump typical of four cylinder 970 and 1070 models. Refer to Fig. 40 for legend except for balancer drive gear (20) and boss (B) for gear.

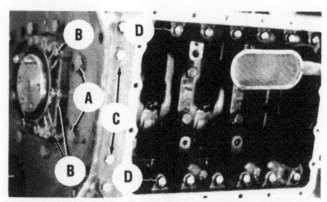

Fig. 42—The crankcase extension is attached to the cylinder block (crankcase) with screws (C & D). Extension is attached to the engine rear plate and rear seal housing with screws (A & B).

CRANKSHAFT FRONT OIL SEAL

770, 870, 970 And 1070 Models

101. The crankshaft front oil seal (23—Fig. 39) is located in timing gear cover and can be renewed after removing crankshaft pulley (2—Fig. 40 or Fig. 41). Pry seal from timing cover taking care not to damage cover. Seal lip rides on hub of crankshaft pulley. Inspect pulley hub for excessive wear at point of seal contact. Install new seal with lip toward inside using suitable driver. Lubricate seal lip and crankshaft pulley hub, install pulley on crankshaft and tighten pulley retaining screw. The crankshaft pulley of some models is equipped with a pin which must engage a hole in crankshaft gear. Coat threads of center screw (3) with medium strength "Loctite" or equivalent and tighten to 115 N•m (85 ft.-lbs.) torque.

CRANKSHAFT REAR OIL SEAL

770, 870, 970 And 1070 Models

102. The crankshaft rear oil seal (14—Fig. 40 or Fig. 41) can be renewed after removing flywheel and seal retainer housing (13). Install new oil seal with lip toward inside. Press seal into housing until flush with outer surface of retainer housing. If crankshaft is grooved at original oil seal contact surface, new seal can be installed 3 mm (⅛ inch) farther into seal housing. Lubricate seal lip and install retainer housing. If not equipped with gasket between seal housing and crankcase, coat sealing surface of retainer housing with RTV sealer. Tighten screws (B—Fig. 42) to 49 N•m (36 ft.-lbs.) torque and screws (A) to 27 N•m (20 ft.-lbs.) torque.

FLYWHEEL

770, 870, 970 And 1070 Models

103. To remove flywheel, first remove engine as outlined in paragraph 80 or 81. Unbolt and remove clutch assembly, then unbolt and remove flywheel from rear of crankshaft. The engine rear plate (16—Fig. 40 or Fig. 41) can be unbolted and removed from rear of engine block after removing flywheel.

When assembling, tighten screws attaching engine rear plate (16) to rear of engine to 49 N•m (36 ft.-lbs.) torque and screws attaching flywheel (18) to crankshaft to 83 N•m (61 ft.-lbs.) torque. Install and tighten screws or nuts to the following torque values: M8 to 26 N•m (230 in.-lbs.); M10 to 49 N•m (36 ft.-lbs.); M12 (nut) to 88 N•m (65 ft.-lbs.).

OIL SUMP

770, 870, 970 And 1070 Models

104. Drain engine oil, then refer to paragraph 80 or 81 and remove engine. Unbolt and remove sump cover (21—Fig. 40 or Fig. 41) from engine. Oil pick-up (25) can be removed for cleaning.

Clean and dry sealing surfaces for oil pick-up tube, sump extension (23) and oil sump cover, then install new gasket or coat sealing surface with RTV sealer. Tighten screws retaining oil pick-up and oil sump cover to 26 N•m (230 in.-lbs.) torque.

OIL PUMP AND PRESSURE RELIEF VALVE

770, 870, 970 And 1070 Models

105. The gerotor type oil pump is mounted on front of engine timing gear housing. Drive gear (5—Fig. 38) on pump shaft is driven by crankshaft gear (1).

To remove oil pump, first remove timing gear cover as outlined in paragraph 93, then measure backlash between oil pump gear (5) and crankshaft gear (1). Backlash should be 0.04-0.12 mm (0.0016-0.0047 inch), but should not exceed 0.20 mm (0.008 inch). Unbolt and remove pump.

Pump end cover (7—Fig. 39) can be removed from pump body (11) to inspect pump. End clearance of rotors in pump body should not exceed 0.13 mm (0.005 inch). Clearance between outer rotor (10) and pump housing should not exceed 0.25 mm (0.10 inch). Clearance between tip of inner rotor (9) and center of outer rotor lobe should not exceed 0.15 mm (0.006 inch). Pump drive gear (5) can be pressed from shaft (8) if renewal is required, but service pump is available with drive gear. Mark on side of outer rotor (10) should be toward inside of pump housing bore.

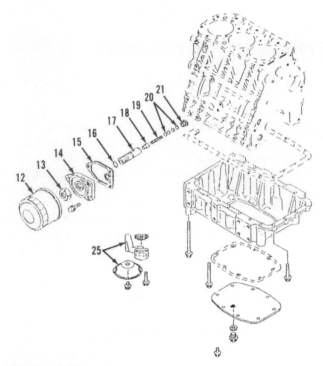

Fig. 43—Exploded view of the engine oil pressure relief valve and filter.

12. Filter
13. Nut
14. Adapter housing
15. Gasket
16. "O" ring
17. Adapter

18. Relief valve poppet
19. Spring
20. Adjusting shims
21. Valve retainer
25. Oil pump pick-up & screen

Install pump using new gasket and tighten retaining screws to 25 N·m (18 ft.-lbs.) torque. Refer to paragraph 93 for installing timing gear cover.

The oil pressure relief valve (18 through 21—Fig. 43) is located inside threaded oil filter adapter. Relief valve pressure can be adjusted by removing valve retainer (21) and adding shims (20). Adding or removing one 1 mm (0.039 inch) thick shim will change oil pressure about 15.6 kPa (2.3 psi). Free length of spring (19) is 46.0 mm (1.81 inch) and spring should exert 20.5 N (4.6 lbs.) when compressed to 27.5 mm (1.08 inch). Relief valve poppet (18) should move freely in bore of body (17) and should not stick. Screws attaching adapter housing (14) to engine should be tightened to 27 N·m (20 ft.-lbs.) torque. If nut (13) is loosened, tighten to 30 N·m (22 ft.-lbs.) torque.

ENGINE BALANCER

970 And 1070 Models

106. The engine balancer consists of two weighted shafts that rotate in opposite directions at twice crankshaft speed. If properly timed, the balancer weights will be positioned at their lowest point (flat

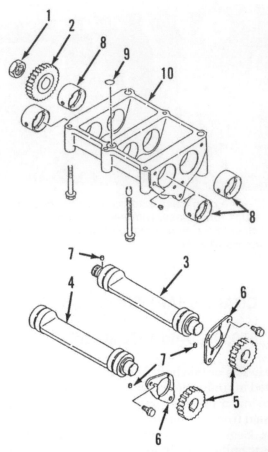

Fig. 44—Exploded view of balancer assembly used on four cylinder models.

1. Nut
2. Driven gear
3. Balance shaft
4. Balance shaft
5. Gears

6. Thrust bearings
7. Drive pins
8. Bushings
9. "O" ring
10. Balancer frame

surfaces will be toward top) each time the pistons are at TDC and BDC of their stroke.

The balancer unit is driven by the balancer drive gear (20—Fig. 41) which is located near the rear of crankshaft. Gear (20) meshes with gear (2—Fig. 44) and balance shaft (4) is driven by gears (5) at front of balancer.

107. REMOVE AND REINSTALL. The balancer assembly can be removed after draining oil and removing the oil sump cover (21—Fig. 41). The front-wheel driveshaft should first be removed from models so equipped. Before unbolting and removing balancer, turn engine crankshaft until timing marks (T—Fig. 45) can be seen.

When installing balancer, install new "O" ring (9—Fig. 44). Align marks on gears (20—Fig. 41 and 2—Fig. 44) as shown in Fig. 45. Tighten screws attaching balancer and screws attaching oil sump cover to 27 N·m (20 ft.-lbs.) torque.

Fig. 45—Timing marks (T) on crankshaft gear and balancer drive gear are at rear and should be aligned as shown. Marks can be seen without removing the crankcase extension and sump (23—Fig. 41), but it is removed in illustration for clarity.

108. OVERHAUL. Balancer assembly can be disassembled after removing nut (1—Fig. 44) and gear (2). Shafts (3 and 4) can be removed with gears (5)

Fig. 46—Timing marks on balancer gears (5) must be aligned as shown.

after unbolting thrust plates (6). Gears (5) can be pressed from shafts to remove thrust plates (6).

Marked side of teeth should be toward outside as shown in Fig. 46 and gears (5—Fig. 44) should be pressed onto shafts until flush with the beveled edge of shaft. If new bushings (8) are installed, align oil holes in bushings with passages in bores of balancer frame (10). Lubricate bushings before installing balance shafts and make sure that marks on gears (5) are aligned as shown (Fig. 46). Tighten screws attaching thrust bearings (6—Fig. 44) to 27 N•m (20 ft.-lbs.) torque. Tighten nut (1) to 106 N•m (78 ft.-lbs.) torque.

DIESEL FUEL SYSTEM

109. Model 670 is equipped with a three piston diesel injection pump which is driven by a separate fuel injection camshaft located in the timing gear housing. The pump is controlled by the governor attached to timing gear housing at rear of injection pump camshaft. Nozzles inject fuel into precombustion chambers located in cylinder head.

Models 770, 870, 970 and 1070 are equipped with a three or four piston diesel injection pump attached to the timing gear housing on right side of engine. Pump governor is integral with pump assembly. Nozzles inject fuel directly into the engine combustion chamber.

FUEL FILTERS AND LINES

All Models

110. OPERATION AND MAINTENANCE. The use of good quality, approved fuel, the careful storage and proper handling of fuel for the diesel engine can not be over emphasized.

Because of extremely close tolerances and precise requirements of all diesel components, it is of utmost importance that clean fuel and careful maintenance be practiced at all times. Unless service personnel have been trained to adjust, disassemble and repair the specific injection system and necessary special tools are available, limit service to pump and nozzles to removal, installation and exchange of complete assemblies. It is impossible to recalibrate an injection pump or reset an injector without proper specifications, equipment and training.

A renewable paper filter element type fuel filter is located between fuel feed pump and fuel injection pump. The fuel filter housing incorporates a fuel shut-off valve on some models, but not others. Check the clear filter sediment bowl daily for evidence of water or sediment. Bowl should be drained and new filter installed at least every 200 hours of operation or if filter appears dirty. It may be necessary to bleed air from filter and fuel system as outlined in paragraph 111 when filter is removed.

Fig. 47—Pump inlet fitting (A) and compression fittings to nozzles (B) are shown for 970 model. Other models are similar.

The fuel system should also be bled if the fuel tank has been allowed to run dry, if components within the system have been disconnected or removed, or if the engine has not been operated for a long time. If the engine starts, then stops, the cause could be air in the system. Refer to paragraph 111 for bleeding procedure

111. BLEEDING. Before attempting to bleed the fuel system, make sure that sufficient supply of clean fuel is contained in the tractor fuel tank. The manufacturer suggests that the system will usually be self bleeding, but the following procedure may be required.

Loosen fitting (A—Fig. 47) and open fuel shut-off valve located on filter housing. Fuel should begin to

flow from the loosened fitting. It may be necessary to crank engine with starter to pump fuel. Tighten fitting when fuel without air bubbles run from the loosened connection.

If engine will not start after bleeding at fitting (A) on pump, it may be necessary to advance throttle slightly, loosen fuel line compression fittings at each injection nozzle, then crank engine with the starter to remove air from high pressure lines and nozzles. Tighten compression fitting when fuel without air begins to flow from loosened connection.

INJECTION PUMP

112. Only trained service personnel with proper tools and equipment should attempt to disassemble or adjusted the fuel injection pump. Specific procedures must be performed under conditions of strict cleanliness and exact measurements require accurate test equipment.

670 Model

113. TIMING TO ENGINE. Timing of the injection pump camshaft and drive gear (4—Fig. 29) is considered correct if marks (T1 and T2) on crankshaft gear (1), idler gear (2) and injection pump camshaft gear (4) are aligned as shown. Specific injection timing can be changed by increasing or decreasing thickness of shims (37—Fig. 48). The old shims should not be reinstalled. Always install pump using new shims of same thickness as shims that were originally in-

Fig. 48—View of some injection pump components typical of 670 models.

1. Timing gear cover
4. Injection pump drive gear
21. "O" ring
22. Tachometer gear unit
24. Ball bearing
25. Retainer
26. Drive key
27. Injection pump camshaft
28. Ball bearing
29. Timing gear housing
35. Cover
36. Diesel injection pump
37. Shims
38. Bearing retaining screw

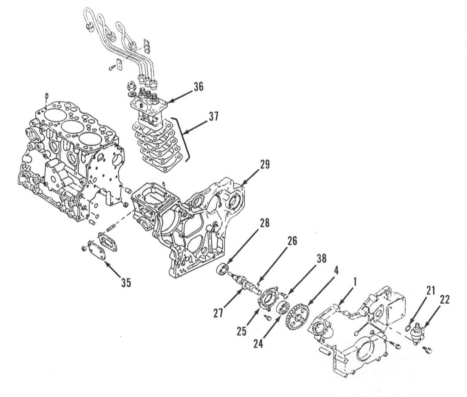

Fig. 49—Pin (A) holds link (B) on pump pin.

stalled. Thickness of shims should not be changed except by specially trained personnel. If original thickness is not known, manufacturer suggests that one new 0.5 mm (0.020 inch) thick shim be installed. Make sure that old paint and sealer are cleaned from mating surface of pump and timing gear housing. Refer to paragraph 114 for removal of pump.

114. REMOVE AND REINSTALL. To remove the injection pump, clean pump and area around pump and lines. Disconnect fuel inlet line to pump, loosen fittings at both ends of high pressure discharge lines between pump and nozzles, then remove high pressure lines. Cover all openings in pump, nozzles and fuel lines to prevent the entrance of dirt. Detach solenoid linkage by removing cotter pin, remove the four screws attaching cover (35—Fig. 48) and solenoid to side of timing gear housing, then remove cover and solenoid. Remove pin (A—Fig. 49), then detach link (B) from pump pin. Remove the four nuts attaching pump to top of timing gear housing and withdraw pump.

Old shims (37—Fig. 48) should not be reinstalled, but new shims should be same thickness as originally installed. Thickness of shims should only be changed by specially trained personnel with proper special tools. Injection timing will occur later if thickness is increased. If original thickness is not known, manufacturer suggests that one new 0.5 mm (0.020 inch) thick shim be installed. Make sure that old paint and sealer are cleaned from mating surface of pump and timing gear housing.

Further disassembly of the pump is not recommended.

When installing, reverse removal procedure, using new shims (37) of same thickness as removed. If original thickness is not known, manufacturer suggests that one new 0.5 mm (0.020 inch) thick shim be installed. Tighten the four nuts retaining pump to 20 N·m (180 in.-lbs.) torque. Compression fittings of high pressure fuel lines should be tightened to 40 N·m (30 ft.-lbs.) torque. Bleed air from fuel system as outlined in paragraph 111.

770, 870, 970 And 1070 Models

115. TIMING TO ENGINE. Timing of the injection pump camshaft and drive gear (4—Fig. 38) is considered correct if marks (T1 and T2) are aligned as shown. Specific injection timing can be changed by moving injection pump in the three elongated mounting slots, but timing should not be changed except by specially trained personnel. Original setting is as shown in Fig. 50. Tighten pump mounting nuts to 26 N·m (19 ft.-lbs.) torque.

116. REMOVE AND REINSTALL. To remove injection pump, clean pump and area around pump and lines. Remove banjo bolts and remove lubrication line (A—Fig. 50) to injection pump. Turn fuel off at shutoff valve of filter housing, then disconnect fuel inlet line to pump and high pressure discharge lines to nozzles from all models. Cover all openings in pump, nozzles and fuel lines to prevent entrance of dirt. Remove small cover from front of timing gear cover, then turn crankshaft until marked tooth valley of injection pump gear is positioned as shown at (T2—Fig. 51).

NOTE: Marked tooth of idler gear may not be aligned with mark on injection pump gear, because idler has uneven number of teeth.

Mark tooth of idler gear, which is meshed with timing mark on pump gear, with paint or chalk as shown to facilitate alignment, then remove nut (39).

NOTE: Do not mark tooth of idler gear with punch or in another permanent method, because gear already has three timing marks (T1, T2 and T3—Fig. 38).

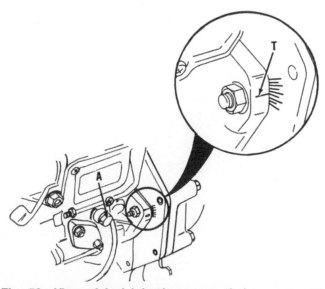

Fig. 50—View of fuel injection pump timing marks (T) typical of 770, 870, 970 and 1070 models. Lubrication line to pump is shown at (A).

Fig. 51—View of timing marks (T2) aligned on 770, 870, 970 and 1070 models. Because of uneven number of teeth on idler gear, mark on idler gear does not align with mark on pump gear every time.

Note position of pump timing marks (T—Fig. 50) for correct reassembly. Remove the three nuts (40—Fig. 52) attaching pump to timing gear housing, then use a suitable gear puller to push pump shaft from gear (4).

NOTE: Do not crank engine with injection pump removed and gear loose in timing gear housing.

If temporary timing marks were not installed on idler gear as described, refer to paragraph 93 and remove timing gear cover. Refer to paragraph 89 and Fig. 38 for aligning timing marks on timing gears.

When installing, reverse removal procedure, making sure that timing marks (T2—Fig. 51) on gears remain aligned. Align injection pump timing marks

(T—Fig. 50) in same position as when removed. Tighten the three pump mounting nuts (40—Fig. 52) to 27 N.m (20 ft.-lbs.) torque and gear retaining nut (39) to 88 N.m (65 ft.-lbs.) torque. Compression fittings of high pressure fuel lines should be tightened using two wrenches. Bleed air from fuel system as outlined in paragraph 111.

GOVERNOR AND INJECTION PUMP CAMSHAFT

670 Model

117. REMOVE AND REINSTALL. To remove the governor (fuel control) and linkage, clean pump and area around pump and governor. Detach solenoid linkage by removing cotter pin, remove the four screws attaching cover (35—Fig. 48) and solenoid to side of timing gear housing, then remove cover and solenoid. Remove pin (A—Fig. 49), then detach link (B) from pump pin. Unbolt and remove engine oil dip stick tube and disconnect speed control cable from lever (13—Fig. 53). Unbolt and remove bracket (1), remove four additional screws attaching governor housing (3) to rear of timing gear housing, then withdraw housing (3).

Inside diameter of sleeve (4) should not be worn larger than 8.20 mm (0.323 inch) and shaft OD should be no less than 7.90 mm (0.311 inch).

Remove nut (5), then remove flyweight assembly (6, 7 and 8). Further disassembly of governor assembly should only be accomplished by trained personnel. Minimum diameter of shaft (16) is 7.90 mm (0.311

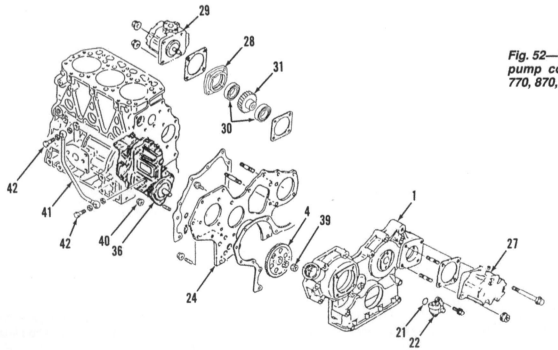

Fig. 52—View of some injection pump components typical of 770, 870, 970 and 1070 models.

1. Timing gear cover
4. Injection pump drive gear
21. "O" ring
22. Tachometer gear unit
24. Timing gear housing
27. Auxiliary pump
28. Adapter
29. Hydraulic pump
30. Bearings
31. Hydraulic pump drive gear
36. Injection pump
39. Nut
40. Nut (3 used)
41. Oil line
42. Banjo bolts

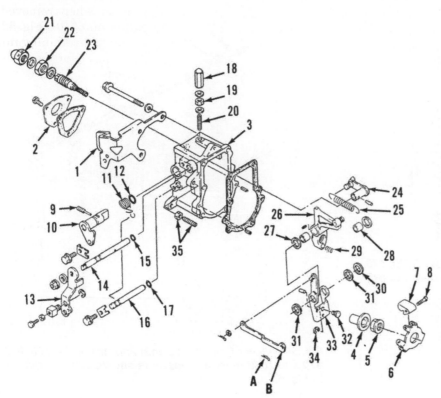

Fig. 53—Exploded view of governor (fuel control) assembly typical of 670 model.

A. Pin	19. Locknut
B. Link	20. High speed
1. Bracket	stop screw
2. Cover	21. Cap nut
3. Housing	22. Locknut
4. Sleeve	23. Fuel control
5. Nut	restrictor
6. Weight retainer	24. Lever
7. Weight (3 used)	25. Spring
8. Pin	26. Lever
9. Retaining screw	27. Shim (2 used)
10. Fuel shut-off lever	28. Bushing (2 used)
11. Return spring	29. Spring
12. "O" ring	30. Shim
13. Speed control lever	31. Washers
14. Shaft	32. Shifter
15. "O" ring (same as 17)	33. Lever
16. Pivot shaft	34. "E" ring
17. "O" ring (same as 15)	35. Idle stop screw
18. Cap nut	& locknut

inch) and bushings (28) should not be worn larger than 8.15 mm (0.321 inch).

To remove the injection pump camshaft (27—Fig. 48), refer to preceding paragraphs and remove governor housing and weights. Refer to paragraph 62 and remove timing gear cover. Refer to paragraph 114 and remove injection pump. Remove idler gear. Working through one of the holes in camshaft drive gear, remove screw (38—Fig. 48) retaining front bearing (24). Tap rear end of camshaft with a soft hammer to remove camshaft (27), bearing (24) and drive gear (4) from timing gear housing (29) and bearing plate (25).

Install new injection pump camshaft if lobe height is less than 30.9 mm (1.217 inches).

When assembling, reverse removal procedure. Refer to Fig. 29 and paragraph 57 for aligning marks on all timing gears, including the injection pump drive gear. Tighten screw (38—Fig. 48) to 20 N·m (180 in.-lbs.) torque. Refer to paragraph 63 to install timing gear housing, paragraph 114 to install injection pump, and to paragraph 118 for adjusting governor.

118. ENGINE SPEED AND GOVERNOR ADJUSTMENTS. Slow idle speed is adjusted by stop screw (35—Fig. 53) and high idle speed by stop screw (20). Settings of the high speed stop (20) and fuel restrictor adjustment (23) are sealed and should be changed only by trained personnel with proper equipment.

Before adjusting low or high idle engine speed stops, make sure that throttle control cable is cor-

rectly attached to lever (13) so that full movement is possible. Low idle speed should be 900-950 rpm and high idle no-load speed should be 3000-3050 rpm.

770, 870, 970 And 1070 Models

119. The governor and injection pump camshaft are integral with the pump and disassembly is not recommended.

120. ENGINE SPEED AND GOVERNOR ADJUSTMENTS. Slow idle speed is adjusted by stop screw (35—Fig. 54) and high idle speed by stop screw (20). Setting of the high speed stop (20) is sealed and should be changed only by trained personnel with proper equipment.

Before adjusting low or high idle engine speed stops, make sure that throttle control cable is correctly attached to lever (13) so that full movement is possible. High idle no-load speed should be 3000-3050 rpm for 770 model; 2750-2800 rpm for 870 and 970 models; 2850-2900 rpm for 1070 model. Low idle speed should be 900-950 rpm for all models.

INJECTOR NOZZLES

All Models

121. TESTING AND LOCATING A FAULTY NOZZLE. If one or more of the cylinders is misfiring and the fuel system is suspected as the cause of

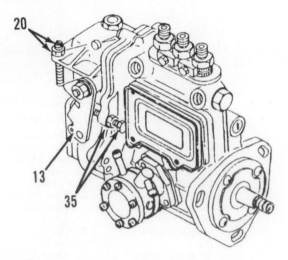

Fig. 54—View of fuel injection pump, typical of 770, 870, 970 and 1070 models, showing governor (fuel control) adjustment points.

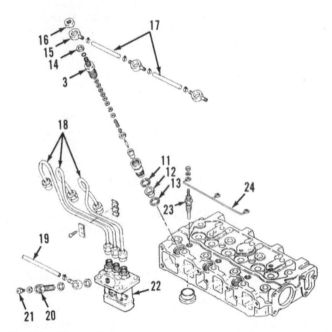

Fig. 55—View of injection pump and nozzle for 670 model. Nozzle is shown exploded at "A" in Fig. 57.

3. Nozzle holder	
11. Gasket	18. High pressure lines
12. Seal	19. Supply line
13. Gasket	20. Fitting
14. Gasket	21. Bleeder fitting
15. Banjo fitting	22. Injection pump
16. Nut	23. Glow plug
17. Return lines	24. Connector

trouble, check by loosening each injector line connection in turn, while the engine is running at slow idle speed. If the engine is not materially affected when an injector line is loosened, that cylinder is misfiring. Remove and test the nozzle or install a new or reconditioned unit as outlined in the appropriate following paragraphs.

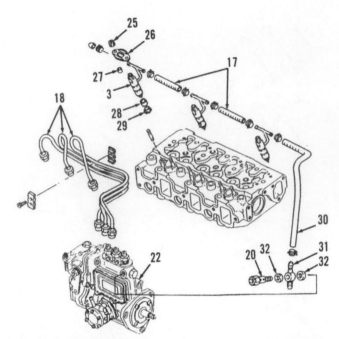

Fig. 56—View of injection pump and nozzle for 770, 870, 970 and 1070 models. Nozzle is shown exploded at "B" in Fig. 57.

3. Nozzle holder	
17. Return lines	27. Spacer
18. High pressure lines	28. Seal
20. Fitting	29. Seat
22. Injection pump	30. Return line
25. Nut	31. Banjo fitting
26. Clamp	32. Gaskets

122. REMOVE AND REINSTALL. Thoroughly clean injector, lines and surrounding area before loosening lines or removing injector. Detach compression fittings at both ends of high pressure lines, using two wrenches, then remove the lines. Disconnect or remove nozzle bleed (return) line from the nozzle or nozzles to be removed and cover all openings to prevent entrance of dirt.

On 670 model, unscrew injector nozzle from cylinder head. Remove injector, gaskets (11 and 13—Fig. 55) and seal (12) from injector bores in cylinder head.

On 770, 870, 970 and 1070 models, remove nuts (25—Fig. 56), holder (26) and spacers (27), then pull injector from cylinder head. Remove seal (28) and seat (29) from bore in cylinder head. If nozzle is difficult to remove from cylinder head, it may be necessary to use a puller attached to the threads for the high pressure fuel line. **Be careful not to damage nozzle while attempting to remove.**

Before installing injectors on all models, first be sure that bore in cylinder head is clean and all seals are removed. Clean the bore in cylinder head carefully and completely, then inspect bore for both cleanliness and damage. Insert new gaskets and seals, then install injector in bore.

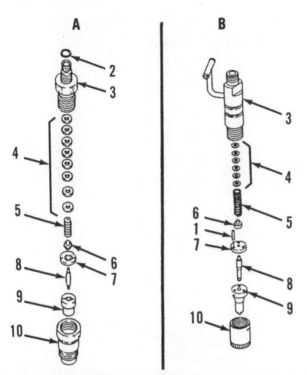

Fig. 57—Indirect injection, pintle type nozzle (A) is used in 670 model; direct injection nozzle (B) is used in 770, 870, 970 and 1070 models.

1. Alignment pins
2. "O" ring
3. Nozzle holder
4. Shims
5. Spring
6. Spring seat
7. Stop plate
8. Nozzle needle
9. Nozzle body
10. Nozzle nut

On 670 model, tighten injector to 50 N·m (37 ft.-lbs.) torque in cylinder head and tighten nut (16—Fig. 55) to 40 N·m (30 ft.-lbs.) torque.

On 770, 870, 970 and 1070 models, tighten injector hold down nuts (25—Fig. 56) to 4.5 N·m (39 in.-lbs.) torque in cylinder head.

On all models, refer to paragraph 111 and bleed the fuel system. Bleed high pressure lines before tightening line fittings to injectors.

123. TESTING. A complete job of testing, cleaning and adjusting the fuel injector requires removal as outlined in paragraph 122 and the use of special test equipment. Use only clean, approved testing oil in the tester tank. Injector should be tested for opening pressure, seat leakage and spray pattern. Before connecting the injector to the test stand, operate tester lever until oil flows, then attach injector to tester line.

WARNING: Fuel leaves injector nozzle with sufficient force to penetrate the skin. Keep exposed portions of your body clear of nozzle spray when testing.

OPENING PRESSURE. Open valve to tester gauge and operate tester lever slowly while observing

gauge reading. Opening pressure should be 11,242-12,202 kPa (1630-1770 psi) for 670 model. Opening pressure should be 19,120-20,080 kPa (2773-2913 psi) for 770, 870, 970 and 1070 models. Opening pressure is changed by adding or removing shims (4—Fig. 57). Set opening pressure to upper limit if opening pressure is changed.

SPRAY PATTERN. Operate the tester lever slowly and notice the spray from the nozzle. Nozzle should emit a chattering sound and should spray a fine stream. Operate tester handle fast and again observe the spray from nozzle. Spray should be finely atomized and the pattern should be even. If nozzle does not chatter, spray is not finely atomized or sprays to one side, disassemble and clean or renew the nozzle.

LEAKAGE TEST. Wipe end of nozzle with a clean cloth after completing opening and spray pattern tests.

On 670 model, operate tester until pressure reaches pressure of 11,032 kPa (1600 psi) and maintain this pressure. Nozzle should not drip for at least 10 seconds; however, end of nozzle may become wet.

On 770, 870, 970 and 1070 models, operate tester until pressure reaches 17,640 kPa (2550 psi) and maintain this pressure. Nozzle should not drip for at least 5 seconds; however, end of nozzle may become wet.

On all models, leakage can usually be corrected by cleaning.

124. OVERHAUL. Do not use hard or sharp tools, emery cloth, grinding compound or other than approved solvents or lapping compounds. Approved nozzle cleaning kits are available from specialized tool sources.

Wipe all dirt and loose carbon from exterior of nozzle and holder assembly. Refer to Fig. 57 for exploded view and proceed as follows:

Secure nozzle in a soft jawed vise or holding fixture and remove nut (10). Place all parts in clean calibrating oil or diesel fuel as they are removed. Use a compartmented pan and exercise care to keep parts from each injector together and separate from other units that are disassembled at the same time.

Clean exterior surfaces with a brass wire brush, soaking parts in an approved carbon solvent if necessary, to loosen hard carbon deposits. Rinse parts in clean diesel fuel or calibrating oil immediately after cleaning to neutralize the solvent and prevent etching of polished surfaces.

Clean nozzle spray hole from inside using a pointed hardwood stick or wood splinter, then scrape carbon from pressure chamber using a hooked scraper. Clean valve seat using brass scraper. Perform the following slide test to check condition of nozzle assembly. Dip nozzle needle (8—Fig. 57) in clean diesel fuel. Insert

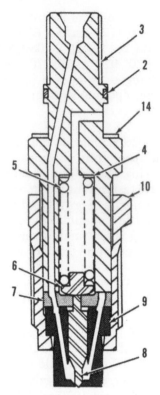

Fig. 58—Cross section of injector used on 670 model. Refer to Fig. 57 for legend.

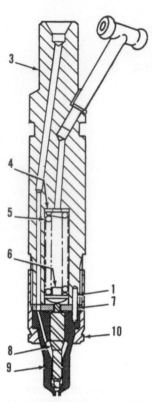

Fig. 59—Cross section of injector used on 770, 870, 970 and 1070 models. Refer to Fig. 57 for legend.

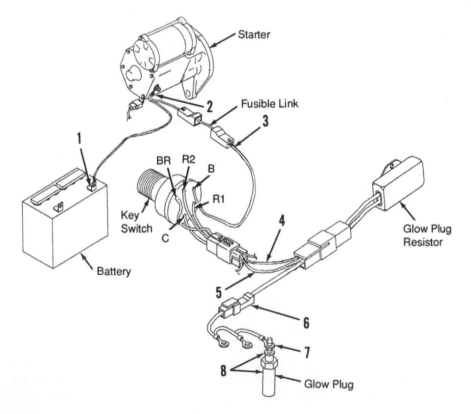

Fig. 60—View of glow plug wiring for 670 model, showing test points. Refer to text for procedure and values.

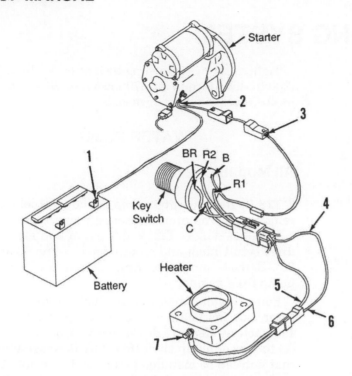

Fig. 61—View of inlet air pre-heater typical of 770, 870, 970 and 1070 models. Refer to text for procedures and values.

valve in nozzle body (9). Withdraw nozzle needle approximately half way from body, then release the needle. The needle should slide to its seat by its own weight. If needle does not slide freely, reclean and check again. Renew nozzle assembly if needle does not slide freely to its seat after cleaning. Refer to Fig. 58 and Fig. 59 for cross-sectional views of injectors.

Reclean all parts, rinsing thoroughly in clean diesel fuel or calibrating oil and assemble while parts are immersed in clean fluid. Make sure adjusting shim pack is intact. Tighten nozzle retaining nut to 40 N·m (30 ft.-lbs.) torque for 670 model. Tighten nozzle retaining nut to 43 N·m (31 ft.-lbs.) torque for 770, 870, 970 and 1070 models. Do not overtighten. Distortion may cause valve to stick and no amount of overtightening can stop a leak caused by scratches or dirt. Retest assembled injector as previously outlined.

PREHEATER / GLOW PLUGS

All Models

125. On 670 model, three glow plugs (23—Fig. 55) are connected in parallel with each glow plug grounding through mounting threads. A preheater is located at the inlet of engine inlet manifold of 770, 870, 970 and 1070 models. On all models, starting (key) switch is provided with "preheating" position which can be used to energize the glow plugs or preheater for faster warm up.

To test 670 model, refer to Fig. 60. Check battery voltage (1) for reference. Battery voltage should be

11.8-13.2 volts. Check voltage between terminal (2) at starter and ground. Voltage at (2) should be the same as battery voltage. System is equipped with a fusible link. Voltage at white/red wire (3) and key switch terminal (B) should be battery voltage. To test key switch, turn switch to "PRE-HEAT" position and check for battery voltage at terminal (R1) and at black/yellow (or black/brown) wire (4). Turn key switch to start position to check for battery voltage at terminal (R2) and at black/red wire (5), connector (6) and front glow plug (7). If a single glow plug is suspected of being damaged, disconnect wire and check resistance (8) between center terminal and ground. Resistance of good glow plug should be 0.2 ohms.

To test 770, 870, 970 and 1070 models, refer to Fig. 61. Check battery voltage (1) for reference. Battery voltage should be 11.8-13.2 volts. Check voltage between terminal (2) at starter and ground. Voltage at (2) should be the same as battery voltage. System is equipped with a fusible link. Voltage at white/red wire (3) and key switch terminal (B) should be battery voltage. To test key switch, turn switch to "PRE-HEAT" position and check for battery voltage at terminal (R1) and at black/yellow (or black/brown) wire (4). Turn key switch to start position to check for battery voltage at terminal (R2) and at black/red wire (5), connector (6) and pre-heater terminal (7). If the pre-heater is suspected of being damaged, disconnect wires from terminal (7) and check resistance from terminal to ground. Resistance of good heater unit should be 0.6 ohms.

COOLING SYSTEM

RADIATOR

All Models

126. Radiator pressure cap is used to increase cooling system pressure. Valve in cap is set to open at about 97-104 kPa (14-15 psi). Cooling system capacity is 3.8 L (4 qt.) for 670 model; 4.8 L (5.0 qt.) for 770 model; 5.2 L (5.5 qt.) for 870 model; 5.8 L (6.1 qt.) for 970 and 1070 models.

To remove radiator from all models, drain coolant, remove grille, engine side panels, hood, and air cleaner inlet hose. Detach upper and lower hoses from radiator. Remove screen from in front of radiator. Detach or remove radiator supports and remove the coolant recovery tank. Detach fan shroud and move to rear over front of engine. Remove the radiator mounting nuts and lift radiator from tractor.

Reassemble by reversing removal procedure. Clearance between fan blades and radiator should be at least 12 mm (½ inch).

THERMOSTAT

All Models

127. The thermostat (18—Fig. 62 or Fig. 63) is located in the coolant outlet and should begin to open at 71° C (160° F). Install new thermostat if not completely open at 85° C (184° F).

Tighten screws attaching coolant outlet to 26 N·m (230 in.-lbs.) torque for 670 model, or 20.3 N·m (180 in.-lbs.) torque for other models.

WATER PUMP

All Models

128. To remove water pump, first drain coolant and remove the radiator. Loosen alternator mounting bolts, then remove fan, water pump and alternator drive belt. Unbolt and remove fan (1—Fig. 62 or Fig. 63). Remove pump mounting screws and separate pump from engine.

Pump repair parts are available from manufacturer, but installation of new pump is often more satisfactory. Measure distance from front face of hub (4) to rear of pump shaft (5) before disassembling so that hub can be installed to correct dimension. Use a puller to remove hub from pump shaft. Unbolt and remove rear plate (11). Support pump housing, then press pump shaft and bearing, seal (8) and impeller (9) rearward from pump housing. Press pump shaft out of impeller.

Press only on outer edge of shaft bearing (5) and press into housing until front of bearing is flush with front of housing. Press seal (8) into housing until it is bottomed in bore, using a tool that contacts only the outer edge of seal.

Support rear end of pump shaft (do not support housing), then press hub (4) onto pump shaft (5) until flush with end of shaft for 670 model or until bottom

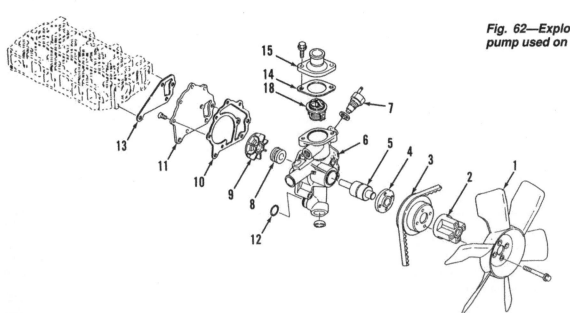

Fig. 62—Exploded view of coolant pump used on 670 model.

1. Fan
2. Spacer
3. Pulley
4. Hub
5. Pump shaft & bearing
6. Pump housing
7. Temperature sensor
8. Seal
9. Impeller
10. Gasket
11. Cover
12. Seal
13. Gasket
14. Gasket
15. Outlet
18. Thermostat

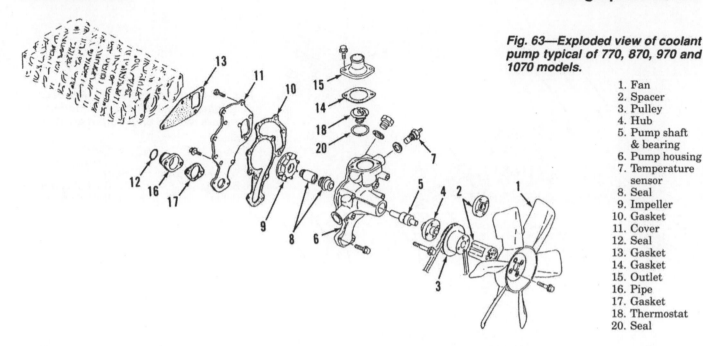

Fig. 63—Exploded view of coolant pump typical of 770, 870, 970 and 1070 models.

1. Fan
2. Spacer
3. Pulley
4. Hub
5. Pump shaft & bearing
6. Pump housing
7. Temperature sensor
8. Seal
9. Impeller
10. Gasket
11. Cover
12. Seal
13. Gasket
14. Gasket
15. Outlet
16. Pipe
17. Gasket
18. Thermostat
20. Seal

of flange is 17 mm (0.67 inch) from top of housing for other models. On all models, install ceramic part of seal (8) in impeller, then press impeller (9) onto shaft until front surface of impeller is 0.3-1.1 mm (0.012-0.043 inch) from pump housing (6). Rear of impeller will be approximately 1-2 mm (0.04-0.08 inch) below flush with rear of housing. Install gasket (10) and rear plate (11), tightening retaining screws to 9 N·m (78 in.-lbs.) torque.

Screws retaining fan (1) and pulley (3) to front hub should be tightened to 11 N·m (96 in.-lbs.) torque.

Screws attaching pump to engine should be tightened to 26 N·m (230 in.-lbs.) torque. Install pump in reverse of removal procedure.

Location, style and thickness of spacer (2) and hub (4) locates fan and pulley. Be sure pulley (3) is aligned with alternator and crankshaft pulleys. Tighten drive belt until approximately 13 mm (½ inch) slack can be measured at midpoint between alternator pulley and crankshaft pulley, when pressing with moderate thumb pressure.

ELECTRICAL SYSTEM

ALTERNATOR AND REGULATOR

129. A 20 amp alternator (Fig. 64) is used on all models. A 35 amp alternator and regulator is available as an option.

All Models

130. TEST. To check no-load output of 20 amp alternator, disconnect the lead wires from the alternator. Place a voltmeter between the two alternator leads; the alternator should develop at least 40 volts of AC when engine is operating at high idle speed. If AC voltage is less than 40 volts, repair or renew alternator. On 20 amp alternator, stop engine and check for continuity between the two alternator leads. Continuity should exist between the two leads

and high resistance would indicate an open circuit. Check for continuity between each of the two alternator leads and alternator frame (ground). Continuity should not exist between either lead and ground. Continuity would indicate a short.

To check no-load output of 35 amp alternator, disconnect the three lead wires from the alternator. Place a voltmeter between one of the three alternator leads and ground, then operate engine at high idle speed. The alternator should develop at least 11 volts of AC. Check the other two leads in the same way. If AC voltage is less than 11 volts with any of the three tests, repair or renew alternator. On 35 amp alternator, stop engine and check for continuity between each of the three leads from alternator and alternator frame (ground). Continuity should exist between the leads and ground. High resistance would indicate an open circuit.

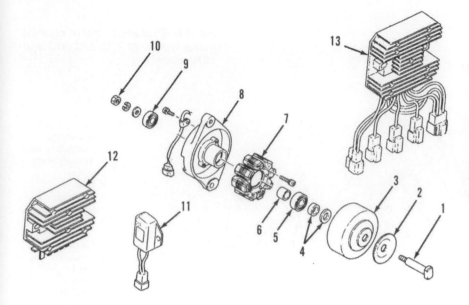

Fig. 64—Exploded view of Kokosan 20 Amp alternator typical of all models. Alternators with 35 amp output are similar.

1. Shaft
2. Pulley half
3. Rotor
4. Washer & spacer
5. Bearing
6. Collar
7. Stator
8. Plate
9. Bearing
10. Nut
11. Glow plug resistor
12. Regulator (20 amp)
13. Regulator (35 amp)

On all models, with the charging circuit fully connected and with a fully charged 12-volt battery, regulated voltage should be 14.2-14.8 volts. If battery load tester is available, run engine at high idle speed and check battery voltage. If less than 14 volts, turn load knob until maximum amperage is obtained. Maximum amperage should be more than 17 amps for 20 amp system and should be more than 32 amps for 35 amp rated system. Perform load test quickly. **Do not apply full load for more than 10 seconds.** If minimum readings are not obtained, renew regulator and retest.

131. OVERHAUL. To disassemble alternator, clamp pulley in a vise and remove nut (10—Fig. 64). Tap end of shaft (1) with a soft hammer to separate flywheel (3) from stator housing (8). Remove clamp from stator leads, remove stator mounting screws and remove stator (7) from housing.

Renew flywheel if magnets are loose or damaged. Check stator for indications of burned wiring or other physical damage and renew as necessary.

To reassemble, reverse disassembly procedure. Tighten nut (10) to 27 N·m (20 ft.-lbs.) torque.

STARTER

All Models

132. All models are equipped with Nippon Denso starters. Removal procedure is conventional. To disassemble, disconnect field lead, remove through-bolts and separate starter motor from solenoid housing. Remove end frame (1—Fig. 65). Pull brush springs away from brushes and lift field coil brushes from brush holder. Pry springs away from ground brushes,

pull brushes up about 6 mm (¼ inch), then release springs so that springs will hold brushes up. Separate brush holder (2) from field coil housing (3). Withdraw armature (5) from field housing.

To disassemble reduction gears and clutch assembly, remove retaining screws and separate solenoid housing (10) from clutch housing (25). Remove clutch assembly and drive pinion from clutch housing. Push retainer (23) back and remove retainer ring (24), then withdraw components from clutch shaft (16).

Remove cover (7) from solenoid housing and withdraw plunger (8). Remove contact plates (9) if necessary. Notice that contact plate located on left side of housing is smaller than contact plate on right side.

Inspect all components for wear or damage. Install new brushes if length is less than 8.5 mm (0.30 inch).

Use an ohmmeter to test for grounded or open winding. Check for shorts between commutator segments and armature shaft. Continuity should exist between two commutator segments. An armature growler should be used to check condition of armature. Field winding can be checked with an ohmmeter. Check for short circuit between field coil brush and field frame. Continuity should exist between the two field brushes.

Reassemble starter by reversing disassembly procedure while observing the following. Grease bearings of clutch assembly (18—Fig. 65). Install spring (17), clutch assembly (18) and washer (19) onto clutch shaft (16), then install retainer (20) and turn retainer to hold parts on shaft. Install spring (21) and drive gear (22), compress parts (16 through 22) in a vise, then install retainer (23) and snap ring (24). Slide retainer (23) over snap ring and release parts from vise. Apply grease to rollers (12), retainer (13) and pinion gear (11), then install parts (11, 12 and 13) on shaft of clutch housing (25) at the same time clutch and shaft assembly (16 through 24) is installed in clutch housing (25). Install steel ball (14) into clutch shaft (16), using grease to hold bearing in place. Install spring (15) in bore of solenoid housing (10) and assemble solenoid housing (10) and clutch housing (25). Attach housings (10 and 25) together with the two screws. Grease felt washer (26) before assembling. Be sure that field coil brush wires do not contact end frame when installing end frame (1). Tighten the two through-bolts to 88 N·m (65 ft.-lbs.) torque.

SAFETY SWITCHES

All Models

133. A safety switch is provided to prevent starting unless the transmission is in neutral. Another safety switch is provided which prevents operation of the pto unless an operator is located on the operator seat. Refer to drawing (Fig. 66) and wiring diagram (Fig. 67) for schematic of the safety circuits.

The neutral start relay is a plug-in module that can be checked separately as follows. Apply battery positive voltage to terminal marked "85" and attach battery ground (negative) to terminal "86." Listen for a click as battery is attached. Check for continuity (closed circuit) between terminals "87" and "30." Also check for open circuit (no continuity) between termi-

nal "87A" and terminal "30." Install new relay if faulty.

The seat safety switch has three positions. In the fully up position, the two terminals on one side of connector should be closed (continuity) and the two terminals on other side of connector should be open (no continuity). Continuity should be reversed when switch is in the fully down position. When switch is in center position, all circuits should be open as indicated by no continuity between any of the four terminals of connector. Install new switch if faulty.

CIRCUIT DESCRIPTION

All Models

134. Refer to Fig. 67 for wiring diagram. A fusible link is located in wire from starter motor solenoid to key switch. The light switch positions are: No terminals are connected when in "OFF" position; "WARN" which connects "BAT" and "WA" terminals; "WORK" which connects "BAT, HETL" and "FL" terminals; "RUN" which connects "BAT" to "HETL" and "WA" terminals. The key switch positions are: "PRE-HEAT" which connects "BAT" and "R1" terminals; no terminals are connected when in "OFF" position; "ON" which connects "BAT" and "BR" terminals; "START" which connects "BAT, C, R2" and "BR" terminals.

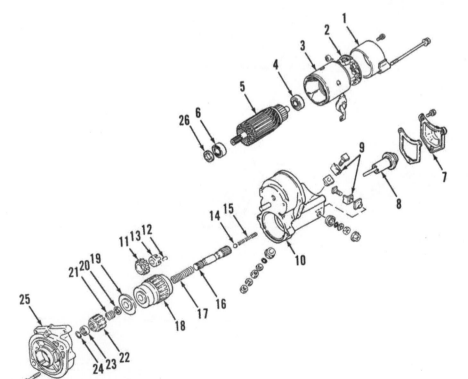

Fig. 65—Exploded view of Nippon Denso starter typical of type used on all models.

1. End frame	14. Ball
2. Brush holder	15. Spring
3. Field coil frame	16. Clutch shaft
4. Bearing	17. Spring
5. Armature	18. Clutch assy.
6. Bearing	19. Washer
7. Cover	20. Toothed retainer
8. Plunger	washer
9. Contact plates	21. Spring
10. Solenoid housing	22. Drive gear
11. Pinion gear	23. Retainer
12. Roller	24. Snap ring
13. Retainer	25. Clutch housing
	26. Felt washer

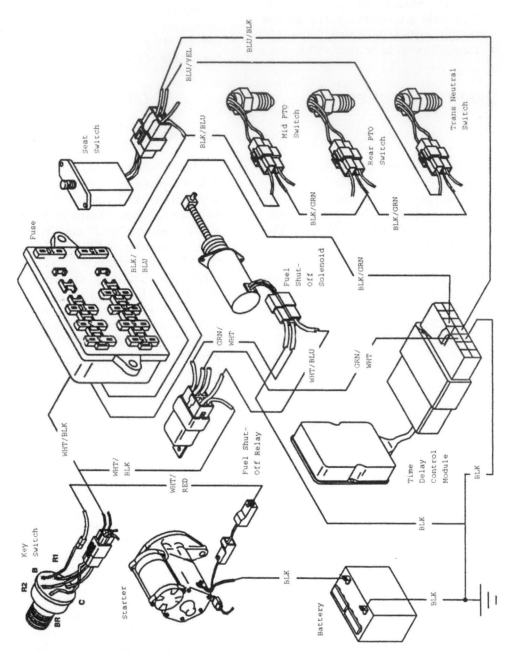

Fig. 66—Drawing of safety switches. Refer to text for tests.

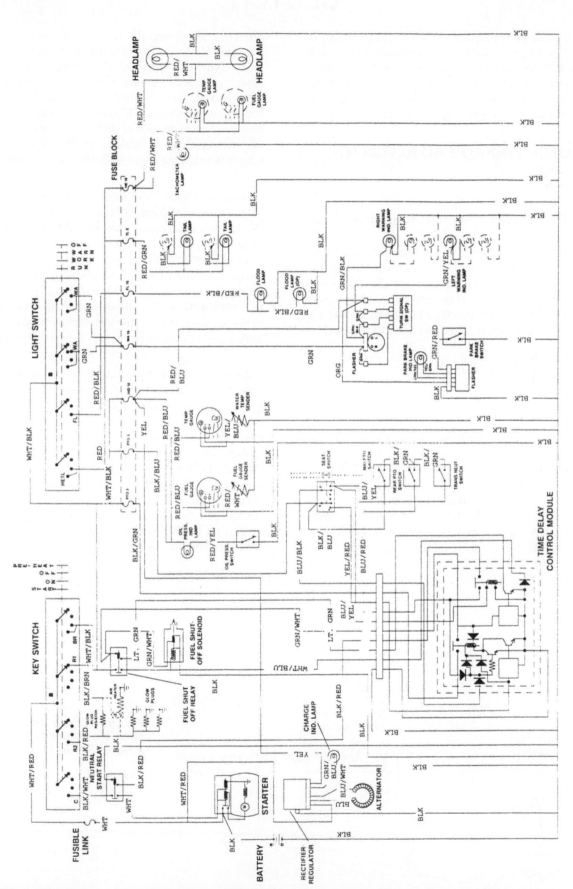

Fig. 67—Wiring schematic typical of all models. Some differences may be noted, if equipped with 35 amp alternator.

CLUTCH

135. All models are equipped with either a dual or single stage engine clutch. Continuous live pto is provided with the dual stage clutch. Control of pto is by engaging or disengaging a separate jaw type pto clutch. Refer to the appropriate following paragraphs for servicing both types of engine clutches.

CLUTCH LINKAGE ADJUSTMENT

All Models

136. ADJUSTMENT. Clutch pedal free travel, measured at top of pedal (F—Fig. 68 or Fig. 69), should be 18-22 mm (¾-⅞ inch). Adjustment is made by changing the length of the clutch operating rod. Loosen locknut, then turn turnbuckle (T—Fig. 68 or Fig. 69) as required to provide correct pedal free play. If clutch disk is new, adjust free travel to 22 mm (⅞ inch). Retighten locknut, then recheck free travel.

CLUTCH SPLIT

All Models

137. To separate tractor between rear of engine and front of clutch housing, remove the hood, engine side shields and lower dash covers. On 670 and 770 models, close fuel shut-off valve at filter, disconnect fuel line from tank to filter, drain fuel from tank, then remove mounting brackets and lift fuel tank from tractor. Remove front-wheel driveshaft from all models so equipped. Disconnect battery and remove en-gine starter from all models. Remove the muffler. Lower rockshaft lift arms and detach hydraulic lines from hydraulic pumps. Disconnect drag link from steering arm of models with manual steering or hydraulic lines from control valve of models with hydrostatic power steering. Cover all hydraulic openings to prevent the entrance of dirt. Disconnect and relocate all wiring, hoses, cables, control rods and braces which would interfere with the separation.

Remove rocker arm cover from engine, attach two lifting eyes to engine, then attach hoist to lifting eyes. Wedge between front axle and frame to prevent tipping. Block rear wheels to prevent rolling and support rear of tractor under clutch housing. Make sure that clutch housing and engine are separately, safely and securely supported. Remove the eight screws attaching engine to clutch housing, shift transmission to neutral, then move the supported engine and front axle forward, away from the clutch housing.

Reassemble by reversing removal procedure. Coat splines of clutch shaft lightly with a multipurpose grease before assembling. It may be necessary to turn flywheel to align clutch splines. On 670 and 770 models, tighten the eight screws attaching engine to clutch housing to 54 N·m (40 ft.-lbs.) torque. On 870, 970 and 1070 models, tighten the eight screws attaching engine to clutch housing to 113 N·m (83 ft.-lbs.) torque. Adjust hood latches to provide 10 mm (0.394 inch) hood clearance. Adjust by turning screws after loosening latch clamp screws. Refer to paragraph 111 for bleeding the fuel system.

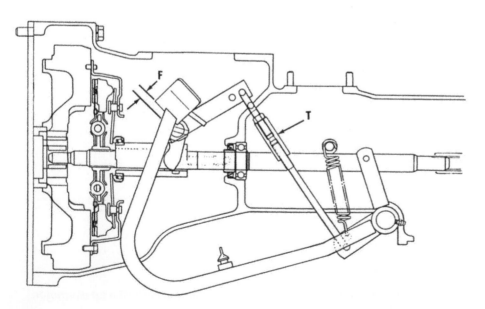

Fig. 68—Cross section of single stage clutch used on some models. Refer to Fig. 69 for dual clutch.

CLUTCH ASSEMBLY

Single Stage Clutch

138. The clutch cover can be unbolted from engine flywheel after separating the engine from the clutch housing as outlined in paragraph 137. Insert a clutch centering tool through clutch disc and into pilot bearing of crankshaft before unbolting the clutch cover to prevent disc from falling.

Check friction disc (11—Fig. 70) for wear by measuring thickness of friction material, which should be 6.4 mm (0.251 in.). Clutch diaphragm release levers should not be worn at the release bearing contact area. Pressure plate and flywheel should be smooth and flat within 0.4 mm (0.016 inch). Measure pilot bushing and install new bushing if inside diameter is more than 15.03 mm (0.592 inch). Use puller to remove bushing if renewal is required.

Remove any oil (including finger prints) from friction surface of flywheel and pressure plate. Light surface rust can be removed using abrasive. Use alignment tool to align and center the clutch friction disc while attaching pressure plate to engine flywheel. Install clutch friction disc with longer hub toward the rear, away from flywheel. Tighten the six retaining screws to 27 N·m (20 ft.-lbs.) torque.

Dual Stage Clutch

139. The clutch cover can be unbolted from engine flywheel after separating the engine from the clutch housing as outlined in paragraph 137. Insert a clutch centering tool through clutch discs and into pilot bearing of crankshaft before unbolting the clutch cover to prevent discs from falling.

The pto drive clutch friction disc is next to the engine flywheel and the traction drive clutch friction disc is located in clutch assembly. Thickness of friction material of disc (11—Fig. 71) should be 7.0 mm (0.270 in.).

On 670 and 770 models, measure pilot bushing and install new bushing if inside diameter is more than 15.03 mm (0.592 inch). Use puller to remove bushing if renewal is required. Flywheel should be smooth and flat within 0.4 mm (0.016 inch).

On 870, 970 and 1070 models, inspect pilot bearing for smoothness and install new bearing if rough or damaged. Use puller to remove bearing if renewal is required. Flywheel should be smooth and flat within 0.24 mm (0.008 inch).

On all models, refer to Fig. 71 when disassembling clutch. It is necessary to remove adjustment screws (15) for the removal of traction drive friction disc (18). Thickness of friction disc (18) material should be 7.6 mm (0.300 inch). Pressure plates (12 and 17) should be smooth and flat within 0.2 mm (0.008 inch).

Remove any oil (including finger prints) from friction surfaces of flywheel and pressure plates. Light surface rust can be removed using abrasive. Use alignment tool to align and center the clutch friction discs while attaching clutch assembly to engine flywheel. Install clutch friction disc with longer hub toward the rear, away from flywheel. Tighten the six retaining screws to 22 N·m (16 ft.-lbs.) torque.

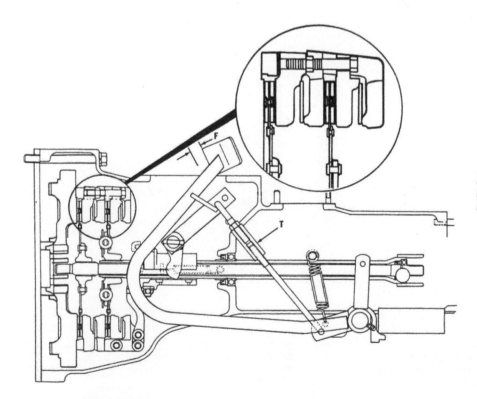

Fig. 69—Cross section of dual clutch used on models with live pto. Refer to Fig. 68 for single stage clutch.

Adjust assembly while installing as follows. Use a feeler gauge to measure distance between head of screws (15) and rear shoulders of pto pressure plate (17). Clearance should be 1.5 mm (0.060 in.) and should be the same for all screws. Tighten locknuts (9) when screws are properly adjusted, then recheck clearance between heads of screws and pressure plate (17). Distance from rear face of adjusting screws (5) and face of flywheel should be 99 mm (3.901 in.) for 670 and 770 models; distance from rear face of screws (5) and flywheel should be 113 mm (4.452 in.) for 870, 970 and 1070 models. Special "U" shaped tool (Fig. 72) is available from manufacturer for measuring distance for 870, 970 and 1070 models. Distance can also be measured by placing straight edge across two of the screws (5), then measuring distance from each end of straight edge to flywheel. It is important that setting of all screws (5) be the same and within 0.7 mm (0.028 in.) of specified setting.

CLUTCH SHAFT AND RELEASE BEARING

All Models

140. Clutch release bearing (1—Fig. 70 or Fig. 71), release fork (25), release shaft (21) and bushings for release shaft can be removed after separating the engine from the clutch housing as outlined in paragraph 137. Be careful not to damage springs (26) when removing or installing. Bushings for cross shaft (21) should be installed 13 mm (0.080 in.) below flush with housing.

On all 670 and 770 models, press front bushing (23—Fig. 70) into bearing sleeve (24) until front of bushing is 13 mm (0.5 in.) from front of sleeve. Press rear bushing (23) into sleeve until the rear of bushing is flush with rear of sleeve.

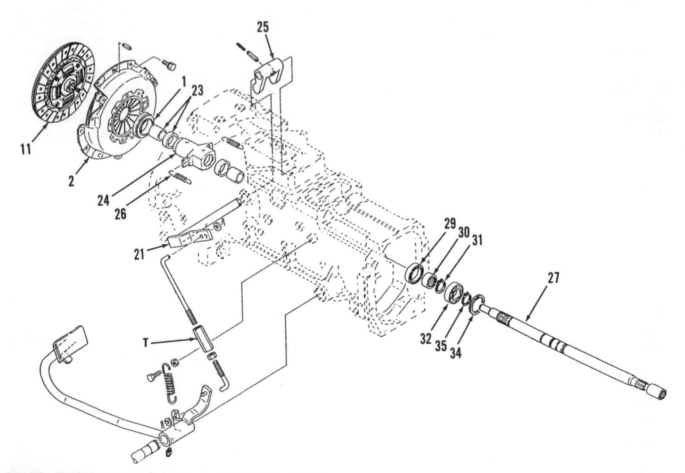

Fig. 70—Exploded view of single stage clutch assembly.

T. Turnbuckle	21. Cross shaft	26. Return spring	31. Snap ring
1. Throwout bearing	23. Bushings & seals	27. Traction drive input shaft	32. Bearing
2. Clutch assembly	24. Throwout bearing sleeve	29. Seal	34. Snap ring
11. Clutch disc (traction drive)	25. Yoke	30. Sleeve	35. Washer

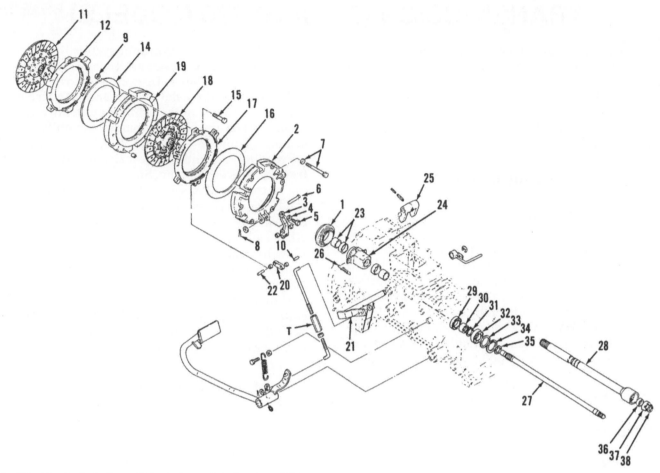

Fig. 71—Exploded view of dual stage clutch assembly.

T. Turnbuckle
1. Throwout bearing
2. Clutch cover
3. Release lever
4. Locknut
5. Adjusting screw
6. Pin
7. Screw & washer
8. Cotter pin
9. Locknut
10. Pin

11. Pto clutch disc
12. Pto pressure plate
14. Pto clutch spring
15. Screw
16. Traction clutch spring
17. Traction drive pressure plate
18. Clutch disc (traction drive)
19. False flywheel for clutch
20. Link

21. Cross shaft
22. Pin
23. Bushings & seals
24. Throwout bearing sleeve
25. Yoke
26. Return spring
27. Pto drive input shaft
28. Traction drive input shaft
29. Seal

30. Sleeve
31. Snap ring
32. Bearing
33. Spacer
34. Snap ring
35. Washer
36. Seal
37. Ball bearing
38. Snap ring

On all 870, 970 and 1070 models, press front bushing (23—Fig. 71) into bearing sleeve (24) until front of bushing is 4.5 mm (0.178 in.) from front of sleeve. Press rear bushing (23) into sleeve until the rear of bushing is flush with rear of sleeve.

To remove input shaft (27—Fig. 70 or 27 and 28—Fig. 72) and front bearing (32—Fig. 70 or Fig. 71), it is necessary to detach clutch housing from both front and rear. Refer to paragraph 137 for separating between engine and clutch housing. Refer to paragraph 143 for separating between clutch housing and transmission housing of 670 and 770 models. Refer to paragraph 150 for separating clutch housing from rear axle center housing of 870, 970 and 1070 models.

Fig. 72—View showing special John Deere Gauge (JDG52) which can be used to measure distance between adjusting screws of clutch fingers to surface of flywheel.

TRANSMISSION (670 AND 770 MODELS)

141. The sliding gear transmission used in 670 and 770 models provides 8 gear reduction ratios forward and 2 reverse ratios. Gear selection is accomplished by positioning the two gear shift levers. Shift patterns are shown in Fig. 73. A safety starting switch is provided and gear selector levers must be in neutral to start engine.

LUBRICATION

All 670 And 770 Models

142. The transmission and hydraulic system share a common reservoir and capacity is 15 L (15.85 quarts). Maintain oil level to full mark on dipstick (D—Fig. 74) located at right rear side of transmission housing. The recommended lubricant is "John Deere Low Viscosity HY-GARD" or equivalent. Transmission and hydraulic system oil should be drained and refilled with new oil every 500 hours of operation.

REMOVE AND REINSTALL

All 670 And 770 Models

143. TRANSMISSION SPLIT. To separate the transmission housing from the clutch housing, first lower the lift arms, then drain hydraulic and transmission lubricating fluid. Remove front driveshaft from models with front-wheel drive. On all models, disconnect battery ground, then remove three point hitch, seat and seat support. Remove hydraulic suction, pressure and return lines which would interfere with separation of transmission and clutch housings.

Cover all openings in hydraulic system to prevent the entrance of dirt. Remove dash covers, clutch housing cover and step plates. Disconnect or remove brake linkage from right pedal to right brake and from cross shaft lever to left brake. Shift main transmission to neutral, then unbolt and remove the shift lever assembly (1 and 2—Fig. 75). Block front wheels to prevent rolling and wedge between front axle and frame to prevent tipping. Support rear of clutch housing so that transmission housing and rear of tractor can be safely rolled back, away from clutch housing. Support the transmission at rear with a floor jack or other suitable method, which will permit transmission and rear of tractor to be safely rolled away from clutch housing. Remove the screws and stud nuts attaching transmission housing to the clutch housing, then carefully move rear of tractor away from clutch housing. When transmission is separated sufficiently from clutch housing, support both halves of tractor with suitable stands to increase safety.

NOTE: If tractor is equipped with dual clutch, the pto rear connecting shaft may remain attached to the clutch and clutch housing and the rear coupling may fall into transmission housing. If coupling falls, it will be necessary to remove the rockshaft housing from top of transmission housing to retrieve and replace coupling.

Rejoin tractor by reversing splitting procedure. Turn rear pto shaft to align shaft splines when moving halves together. Clean mating surfaces of transmission and clutch housing, then apply flexible sealer to both surfaces. Tighten screws and nut attaching housings together to 54 N•m (40 ft.-lbs.) torque. Clean hydraulic suction screen (4) and install new hydraulic filter when assembling and refer to paragraph 142 for lubricant requirements.

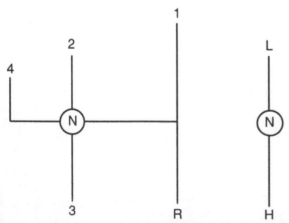

Fig. 73—Gear shifting patterns for 670 and 770 models are shown. Numbers indicate the speed selection of the main transmission and (R) is reverse. The range transmission is provided with low (L) and the high range (H). Neutral positions of both main and range transmission are identified by (N).

Fig. 74—Transmission and hydraulic system fluid should be maintained at level of top mark of dipstick (D). Filler plug is located at top rear of rockshaft housing.

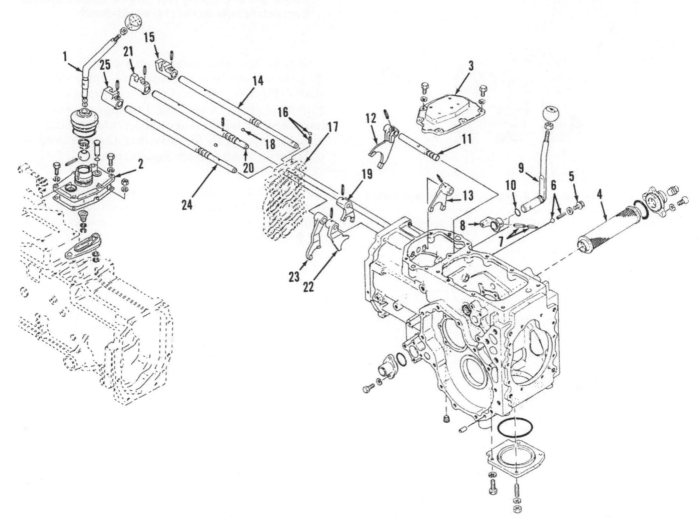

Fig. 75—Exploded view of main and range transmission shifter rails and forks. Detent balls and springs (16) are located in bores of quill under each shift rail. Balls for the three detents (16) and the two interlock balls (18) are all same size.

1. Main shift lever
2. Cover
3. Range cover
4. Hydraulic suction screen
5. Screw
6. Detent spring & ball
7. Roll pins
8. Internal lever
9. Range shift lever
10. "O" ring
11. Range shift rail
12. Range shift fork
13. Fork (4th)
14. Shift rail (4th)
15. Shift gate (4th)
16. Detent ball & spring
    (3 used)
17. Quill
18. Interlock ball
19. Fork (2nd & 3rd)
20. Shift rail (2nd & 3rd)
21. Shift gate (2nd & 3rd)
22. Fork (1st)
23. Fork (Reverse)
24. Shift rail (1st & Rev.)
25. Shift gate (1st & Rev.)

144. R&R RANGE SHIFTER. Remove selective control valve if so equipped and the rockshaft slow return valve of all models, then unbolt and remove cover (3—Fig. 75). Remove screw (5), detent spring and ball (6). Remove roll pins (7), then remove shift lever (9), "O" ring (10) and internal lever (8).

To remove shift rail (11) and range shift fork (12), it is necessary to separate tractor as outlined in paragraph 143 and remove bearing quill (17) as outlined in paragraph 145.

When installing, splits in roll pins (7—Fig. 75) should be 180° apart. Apply medium strength thread lock and sealer to screw (5) to prevent leaking or

loosening. Use sealer on mating surfaces of housing and cover (3).

145. R&R TRANSMISSION. To remove the transmission, first refer to paragraph 143 and split the tractor between the clutch and transmission housings, then remove the range shifter as outlined in paragraph 144. Carefully remove pin from fourth speed shift fork (13—Fig. 75), then withdraw shift rail (14).

NOTE: Be careful to catch roll pin as it is driven from shaft and fork (13). It will be necessary to use a soft drift to drive shaft (14) from shift fork (13). Do not lose ball and detent (16) or interlock ball (18).

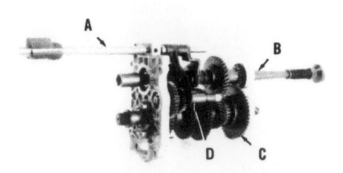

Fig. 76—Transmission gears and shafts are withdrawn with quill.

A. Shift rails
B. Input shaft

C. Bevel pinion shaft
D. Reverse shaft

Detent ball and spring (16) may be quickly ejected from detent hole as rail (14) is withdrawn.

Remove the four screws attaching bearing quill (17), then carefully withdraw quill and transmission shafts (Fig. 76). Be careful to withdraw shafts and quill straight from housing. Pto engagement cam or coupling may fall from rear shaft into housing. Refer to paragraphs 146 for servicing the removed shift rails, forks, shafts, gears and bearings.

When reassembling, apply medium strength thread lock to the four screws attaching quill to transmission housing, then tighten screws to 36 N·m (27 ft.-lbs.) torque.

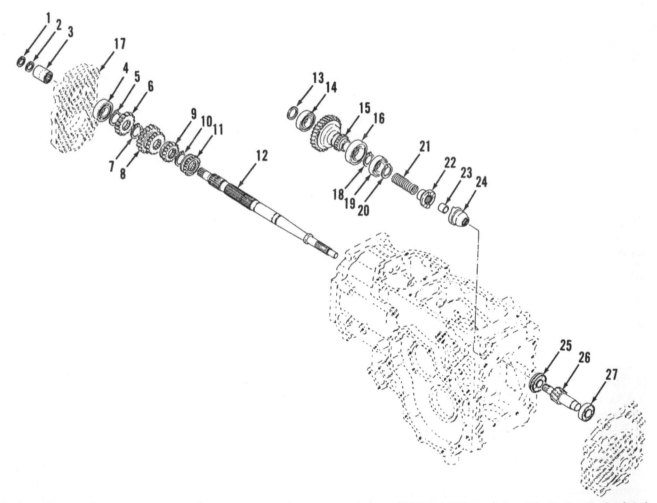

Fig. 77—Exploded view of transmission mainshaft and related parts for 670 and 770 models with single stage clutch.

1. Snap ring	8. Sliding gear	15. Reduction gear	22. Overrunning clutch
2. Snap ring (same as 1)	9. Sliding gear	16. Ball bearing (same as 14)	23. Bushing
3. Coupling	10. Snap ring (same as 5 & 7)	17. Quill	24. Overrunning clutch
4. Ball bearing	11. Sliding gear	18. Snap ring (same as 20)	25. Ball bearing
5. Snap ring (same as 7 & 10)	12. Main (input) shaft	19. Ball bearing	26. Pto upper shaft
6. Gear	13. Spacer	20. Snap ring (same as 18)	27. Ball bearing
7. Snap ring (same as 5 & 10)	14. Ball bearing (same as 16)	21. Spring	

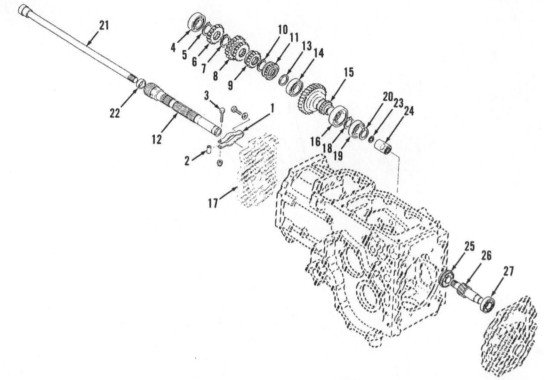

Fig. 78—Exploded view of transmission mainshaft and related parts for 670 and 770 models with dual clutch and live pto.

1. Leaf spring	8. Sliding gear	15. Reduction gear
2. Spacer	9. Sliding gear	16. Ball bearing (same as 14)
3. Bolt & nut	10. Snap ring (same as 5 & 7)	17. Quill
4. Ball bearing	11. Sliding gear	18. Snap ring (same as 20)
5. Snap ring (same as 7 & 10)	12. Main (input) shaft	19. Ball bearing
6. Gear	13. Spacer	20. Snap ring (same as 18)
7. Snap ring (same as 5 & 10)	14. Ball bearing (same as 16)	21. Pto drive shaft

22. Bushing	25. Ball bearing
23. Snap ring	26. Pto upper shaft
24. Coupling	27. Ball bearing

OVERHAUL

All 670 And 770 Models

146. To overhaul the main and range transmissions, refer to paragraph 145 and remove quill, rails, forks, shafts, gears and bearings, then refer to the following paragraphs.

Remove roll pins from shift forks (19, 22 and 23—Fig. 75), then drive shift rails (20 and 24) from forks and carefully withdraw shift rails from bearing quill (17). Catch detent balls and springs (16) and interlocking balls (18). Interlock balls are spring loaded and will be ejected from bores as rails are removed.

Main (input) shaft is different for models with single stage clutch (Fig. 77) and models with dual clutch (Fig. 78). Main shaft, gears and bearings can be bumped from quill, then disassembled as required.

Remove nuts (1 and 2—Fig. 79), washer (3) and spacer (4) or gear (5). Remove the cap screws attaching housing (6) to quill (17), then use two of the screws in threaded holes of housing to push housing away from quill. Be careful not to damage shims (7) or drop shaft. Cluster gear (15) and bearing (13) can be

pressed from bevel pinion shaft (25). Disassembly or removal of remaining parts will be evident.

Reverse shaft (33—Fig. 79) can be removed after removing snap ring (26) and washer (27). Needle bearing (34) is pressed into bore of transmission housing and should be removed only if renewal is required.

Lubricate clean parts, then assemble in reverse of disassembly procedure. Use medium strength "Loctite" or equivalent on threads of nuts (1 and 2) and screws attaching bearing housing (6) to quill (17). Tighten nut (2) to 108 N·m (80 ft.-lbs.) torque, then tighten locknut (1) also to 108 N·m (80 ft.-lbs.) torque. Tighten screws attaching the bearing housing (6) to quill to 36 N·m (27 ft.-lbs.) torque. Unless bevel pinion (25) and ring gear are renewed or mesh position was questioned for some other reason, original thickness of shims (7) can be reinstalled. Refer to paragraph 147 for selecting correct thickness of shims (7) to set bevel gear mesh position.

147. To check mesh position of bevel pinion (25—Fig. 79) with ring gear, paint some teeth of ring gear with machinist dye, then turn bevel pinion until teeth of bevel pinion pass the painted teeth. Observe mesh

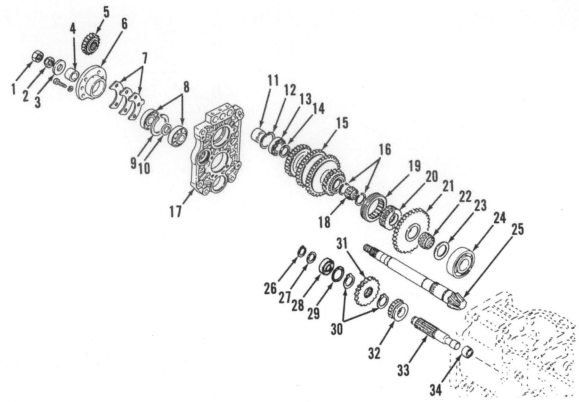

Fig.79—Exploded view of reverse shaft (33), bevel pinion shaft (25) and related parts. Gear (5) is used to drive front wheels of models equipped with front-wheel drive and installed in place of spacer (4).

1. Nut
2. Nut
3. Lock washer
4. Spacer
5. Gear
6. Bearing housing
7. Shims
8. Tapered roller bearings
9. Snap ring
10. Spacer
11. Spacer
12. Snap ring
13. Ball bearing
14. Spacer
15. Cluster gear
16. Snap rings
17. Quill
18. Needle bearing
19. Sliding coupling
20. Sleeve (hub)
21. Gear
22. Needle bearing
23. Thrust washer
24. Ball bearing
25. Bevel pinion shaft
26. Snap ring
27. Washer
28. Ball bearing
29. Snap ring
30. Snap rings
31. Gear
32. Sliding gear
33. Reverse shaft
34. Needle bearing

position as indicated by marks in paint. If mesh position is correct, the teeth of bevel pinion will contact teeth of ring gear in center. Bevel pinion (and mesh position) can be moved toward front by adding shims (7). Bevel pinion teeth will mesh farther to rear

if some shims (7) are removed. Shims (7) move the bevel pinion and some difference in gear backlash may be noted, but backlash should be changed by moving differential and ring gear from side to side as outlined in paragraph 157.

TRANSMISSION (870, 970 AND 1070 MODELS)

148. Collar shift and synchronized shift transmissions are used in 870, 970 and 1070 models. Principles of operation and many service procedures are the same for both collar shift and synchronized models. Both types provide 9 gear reduction ratios forward and 2 reverse ratios. Gear selection is accomplished by positioning the two gear shift levers. Shift patterns are shown in Fig. 80. A safety starting switch is provided and gear selector levers must be in neutral to start engine.

LUBRICATION

All 870, 970 And 1070 Models

149. The transmission and hydraulic system share a common reservoir and capacity is 21 L (22.2 quarts). Maintain oil level to full mark on dipstick (D—Fig. 81) located at rear of transmission housing. The recommended lubricant is "John Deere Low Viscosity

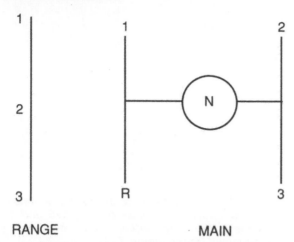

Fig. 80—Gear shifting patterns for 870, 970 and 1070
models are shown. Numbers indicate the speed selection
of the main transmission and (R) is reverse. The range
transmission provides three speed ranges in each of the
forward gear selections and two speed ranges in reverse.
Internal linkage is provided with a block which prevents
operation in high range when main transmission is in
reverse. Neutral position is indicated by (N).

HY-GARD" or equivalent. Transmission and hydraulic system oil should be drained and refilled with new oil every 500 hours of operation.

TRANSMISSION SPLIT

All 870, 970 And 1070 Models

150. The main transmission is located in the rear of clutch housing and the range transmission is located in the front of rear axle center housing. To separate between clutch housing and transmission housing, first lower the lift arms, then drain hydraulic and transmission lubricating fluid. Remove front driveshaft from models with front-wheel drive. On all models, disconnect battery ground, then remove hydraulic suction, pressure and return lines which would interfere with separation of housings. Cover all openings in hydraulic system to prevent the entrance of dirt. Remove rockshaft stop knob, rear pto knob and mid pto knob (if so equipped), then remove cover that surrounds these controls. Remove clutch housing cover and step plates, then unbolt brackets and fenders from step supports or other brackets which prevent separating. Disconnect or remove brake linkage and return springs from right pedal to right brake and from cross shaft lever to left brake. Close fuel shut-off valve at fuel filter, disconnect fuel lines, then cover all fuel system openings. Check around tractor at joint where clutch housing attaches to the rear axle center housing for any wiring, connected springs, hydraulic lines, fuel lines or attached brackets that would interfere with separation. Remove or tie out of the way anything that might be damaged.

Fig. 81—Transmission and hydraulic system fluid for 870,
970 and 1070 should be maintained at level of top mark
on dipstick (D).

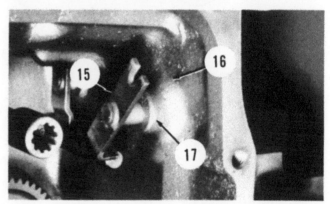

Fig. 82—Lockout arm (15) is located on left side of rear
center housing to prevent operation in reverse while
range transmission is in highest speed.

Shift main transmission to neutral and position the range lever in "2." Block front wheels to prevent rolling and wedge between front axle and frame to prevent tipping. Support rear of clutch housing so that rear of tractor can be safely rolled back, away from clutch housing. Support the rear axle center housing at rear with a floor jack under rear of drawbar or by some other suitable method, which will permit rear of tractor to be safely rolled away from clutch housing. Remove the screws and stud nuts attaching housings together, then carefully move rear of tractor away from clutch housing. When separated sufficiently, support both halves of tractor with suitable stands to improve safety.

> **NOTE: If tractor is equipped with dual clutch, the pto rear connecting shaft may remain attached and the rear coupling may fall into transmission housing. If coupling falls, it will be necessary to remove the rockshaft housing from top to retrieve and replace coupling.**

Rejoin tractor by reversing splitting procedure. Turn rear pto shaft to align shaft splines when moving halves together. Make sure that notch in lockout arm (15—Fig. 82) correctly engages pin (5—Fig. 83)

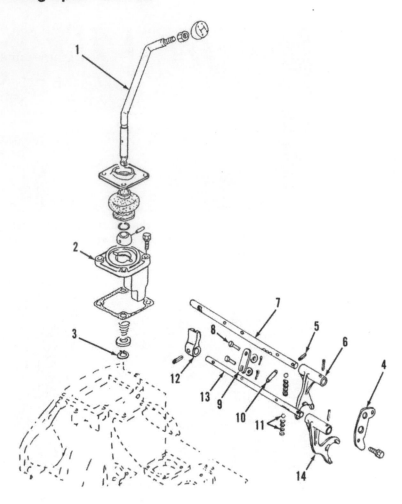

Fig. 83—Exploded view of main transmission shifter assembly typical of 870, 970 and 1070 models. Forks are different than shown for synchronized transmission.

	8. Pins
1. Shift lever	9. Arm
2. Retainer	10. Interlock pin
3. Snap ring	11. Detent ball & spring
4. Lockout plate	(2 used)
5. Pin	12. Shift gate
6. Shift fork (1st & Rev.)	13. Shift rail (lower)
7. Shift rail (top)	14. Shift fork (2nd & 3rd)

in end of shift rail (7). Lockout arm (15—Fig. 82) prevents selecting reverse when range transmission is in highest speed range. Clean mating surfaces of housings, then apply flexible sealer to both surfaces. Tighten screws attaching housings together to 85 N·m (63 ft.-lbs.) torque and nuts to 80 N·m (59 ft.-lbs.) torque. Clean hydraulic suction screen and install new hydraulic filter when assembling and refer to paragraph 159 for lubricant requirements.

MAIN TRANSMISSION

All 870, 970 And 1070 Models

151. R&R SHIFT LEVER, SHIFT FORKS AND RAILS. The shift lever and retainer (1 and 2—Fig. 83) can be unbolted and removed without removing other parts. It is necessary to remove the shafts, gears, shift forks and rails as outlined in paragraph 152, before removing shifter rails (7 and 13) and forks (6 and 14) from the bearing quill. Plate (4) retains interlock pin (10) and is attached to bearing quill. Remove cotter pins, washers and pins (8), then remove arm (9). Remove roll pins from shift forks (6 and 14) and shift gate (12), then drive shift rails (7 and

13) from forks and carefully withdraw shift rails from bearing quill. Catch detent balls and springs (11), which are spring loaded and will be ejected from bores as rails are removed.

Reassemble by reversing procedure. Shift forks for models with synchronized transmission are different than shown in Fig. 83, but are installed in similar way. Refer to paragraph 152 for installing the shafts, gears and quill assembly and to paragraphs 137 and 150 for rejoining transmission housings.

152. R&R AND OVERHAUL. To remove the main transmission shafts and gears (Fig. 84), first separate between the clutch housing and engine as outlined in paragraph 137. Separate clutch housing from the rear axle center housing as outlined in paragraph 150. Unbolt shift lever retainer (2—Fig. 83), then lift lever (1) and retainer assembly from housing. Remove the rear pto shaft, clutch release bearing and neutral start switch. Remove the four screws attaching bearing quill to rear face of clutch housing, tap front of shafts, then withdraw shafts, gears, rails, forks and quill (Fig. 84) from rear.

Refer to Fig. 85 for exploded view of main transmission. Even though some identical parts are installed in more than one location, separate parts as they are

Fig. 84—The main transmission shafts, gears, forks and shift rails are removed from housing positioned in quill as shown. Synchronized transmission is shown, but collar shift is similar.

removed, so that parts can be reinstalled in original location. Individual parts of the synchronizer assemblies are not available and disassembly of these units is not recommended. If synchronizer of transmission so equipped is damaged, a new complete synchronizer should be installed. It is important that oil seal (17) is in good condition and not damaged when assembling. Make sure that all parts, especially bearings, are lubricated with transmission oil as they are assembled. Apply petroleum jelly to washers (23 and 48), then position with grooved sides toward gears (24 and 47).

Assemble shafts, gears, rails, forks and quill (Fig. 84) and insert into housing from rear. Install and tighten the four screws attaching bearing quill to rear face of clutch housing. Install the rear pto shaft,

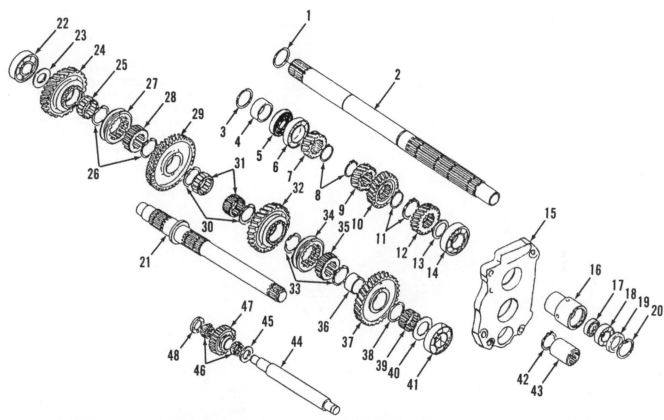

Fig. 85—Exploded view of collar shift main transmission gears and shafts. Except for synchronizer assemblies installed in place of collars (27 and 34) synchronized transmissions are similar.

1. Snap ring (same as 3)	14. Ball bearing (same as 41)	26. Snap rings (same as 33)	38. Snap rings (same as 30)
2. Clutch (input) shaft	15. Bearing quill	27. Shift coupling (same as 34)	39. Needle bearing (same as 25 & 31)
3. Snap ring (same as 1)	16. Oil seal case	28. Hub (same as 35)	40. Washer
4. Locking collar	17. Oil seal	29. Gear (53 teeth)	41. Ball bearing (same as 14)
5. Oil seal	18. Ball bearing	30. Snap rings (same as 38)	42. Snap ring
6. Ball bearing	19. Bushing	31. Needle bearing (same as 25 & 39)	43. Output coupling
7. Gear (21 teeth)	20. Snap ring	32. Gear (44 teeth)	44. Reverse idler shaft
8. Snap rings (same as 11)	21. Output shaft	33. Snap ring (same as 26)	45. Washer
9. Gear (same as 7)	22. Ball bearing	34. Shift coupling (same as 27)	46. Needle bearings
10. Gear (31 teeth)	23. Washer	35. Hub (same as 28)	47. Reverse idler gear (36 teeth)
11. Snap ring (same as 8)	24. Gear (46 teeth)	36. Bearing	48. Washer (grooves toward gear)
12. Gear (27 teeth)	25. Needle bearing (same as 31 & 39)	37. Gear (49 teeth)	
13. Spacer			

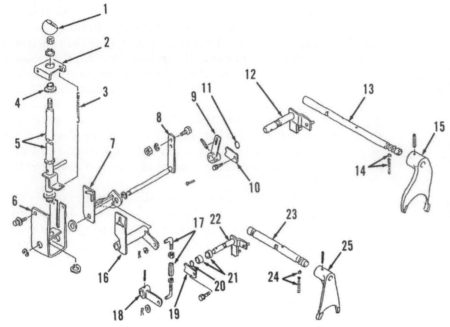

Fig. 86—Exploded view of shift linkage, rails and forks for the range transmission of 870, 970 and 1070 models.

1. Range shift knob	14. Detent ball & spring
2. Support	15. Shift fork
3. Spring	16. Arm
4. Bushing	17. Link
5. Shift lever	18. External lever
6. Support	19. Retainer plate
7. Arm	20. "O" ring
8. Lever	21. Bushings
9. External lever	22. Internal lever
10. Retainer plate	23. Shift rail
11. "O" ring	24. Detent ball & spring
12. Internal lever	25. Shift fork
13. Shift rail	

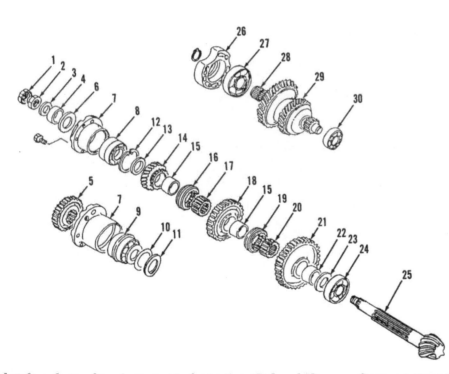

Fig. 87—Exploded view of range transmission shafts and associated parts. Gear (5) is used in place of spacer (4) on models with front-wheel drive. Parts (7, 9, 10 & 11) are different for 1070 models than similar parts (7, 8, 12 and 13) used in other models.

1. Locknut	16. Shift collar (same as 19)
2. Nut	17. Hub (same as 20)
3. Washer	18. Gear (40 teeth)
4. Spacer (2-wheel drive)	19. Shift collar (same as 16)
5. Gear (4-wheel drive)	20. Hub (same as 17)
6. Washer	21. Gear (53 teeth)
7. Retainer	22. Bushing
8. Bearing (870 & 970 models)	23. Washer
9. Bearing (1070 model)	24. Ball bearing
10. Shims	25. Bevel pinion shaft
11. Washer	26. Retainer
12. Snap ring	27. Ball bearing
13. Washer	28. Bearing
14. Gear (5 teeth)	29. Cluster gear assy.
15. Bushings	30. Ball bearing

clutch release bearing, neutral start switch, shift lever (1—Fig. 83) and retainer (2). Attach clutch housing to rear axle center housing as outlined in paragraph 150, then attach clutch housing to engine as outlined in paragraph 137.

RANGE TRANSMISSION

All 870, 970 And 1070 Models

153. R&R SHIFT RAILS AND FORKS. To remove the range transmission shift rails and forks,

first separate between the clutch housing and rear axle center housing as outlined in paragraph 150, then refer to paragraph 192 and remove the rockshaft top cover.

Loosen or remove retainer plates (10 and 19—Fig. 86), then pull levers (12 and 22) out, away from shift rails (13 and 23). Remove roll pins from shift forks (15 and 25), then drive shift rails (13 and 23) from forks and carefully withdraw shift rails. Catch detent balls and springs (14 and 24), which are spring loaded and will be ejected from bores as rails are removed.

Fig. 88—Gear (5) is located on front of bevel pinion shaft if equipped with front-wheel drive. Refer to Fig. 87 for bevel pinion shaft and to Fig. 89 for exploded view of front drive shaft.

Inspect and renew parts as necessary. Reassemble by reversing disassembly procedure. If necessary, adjust length of linkage (17) to align slots in arms (7 and 16) in center mid range position.

154. R&R AND OVERHAUL. To remove the range transmission shafts and gears, first separate between the clutch housing and rear axle center housing as outlined in paragraph 150 and remove the rockshaft top cover as outlined in paragraph 192. Refer to paragraph 153 and remove the shift rails and forks.

Refer to Fig. 87 for exploded view of range transmission. The lower bevel pinion shaft (25—Fig. 87) must be removed before removing the upper shaft. If equipped with front-wheel drive, refer to paragraph 155 to remove the drive gears and shaft located in the center housing. On all models, remove the five screws attaching bearing retainer (7—Fig. 87), then pull the lower (bevel pinion) shaft (25), upper shaft (29) and retainers (7 and 27) forward together.

When disassembling shafts, be careful not to lose or damage shims (10) that adjust mesh position of the bevel pinion. Some parts are identical, but should be kept separate when removed, so that parts can be reinstalled in original location. If equipped with front-wheel drive, gear (5—Fig. 88) should be installed with longer hub toward rear. On all models, tighten nuts (1 and 2—Fig. 87) to 88 N·m (65 ft.-lbs.) torque.

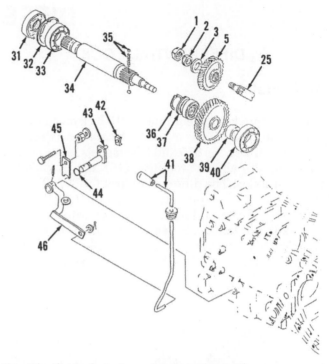

Fig. 89—Exploded view of front-wheel-drive gears and shaft located in front of range transmission.

1. Locknut	36. Snap ring
2. Nut	37. Coupling
3. Washer	38. Driven gear
5. Drive gear	39. Spacer
25. Bevel pinion shaft	40. Ball bearing
31. Oil seal	41. Engagement handle
32. Snap ring	42. Guide
33. Ball bearing	43. Internal lever
34. Output shaft	44. "O" ring
35. Detent balls & spring	45. Retainer
	46. Lever

FRONT-WHEEL DRIVE GEARS

870, 970 And 1070 Models

155. R&R AND OVERHAUL. It is necessary to separate the clutch housing from rear axle center housing as outlined in paragraph 150 to inspect or service gears and output shaft. Seal (31—Fig. 89), snap ring (32) and bearing (33) are located in front bore of center housing. Be careful not to lose detent balls and spring (35). Gear (5) is attached to front of bevel pinion shaft (25) by nuts (1 and 2).

Install gear (5) with longer hub toward rear and tighten nuts (1 and 2) to 88 N·m (65 ft.-lbs.) torque. Clean and dry mating surfaces of housings, then apply flexible sealer to surfaces before assembling.

DIFFERENTIAL

DIFFERENTIAL LOCK

All Models

156. All models are equipped with lock mechanisms which can be engaged to power both rear axles. The differential lock for 670 and 770 models, slides collar (19—Fig. 90) against coupling (16) and couples the two rear axles. The differential lock for 870, 970 and 1070 models, slides pins of locking collar (19—Fig. 91) through differential housing (22) and into right side gear (26) and locks side gears (26 and 28) to housing (22). Refer to paragraphs 159 and 160 for servicing differential lock of 670 and 770 models. Refer to

paragraph 157 for servicing differential lock of 870, 970 and 1070 models.

DIFFERENTIAL AND BEVEL GEARS

All Models

157. R&R AND OVERHAUL. The differential can be removed and serviced without removal of bevel pinion (15—Fig. 90 or Fig. 91), but bevel pinion and bevel ring gear (15 and 20) are available only as a matched set. If necessary to remove the bevel pinion, refer to paragraphs 146 and 154 for removal of the range transmission and bevel pinion (15).

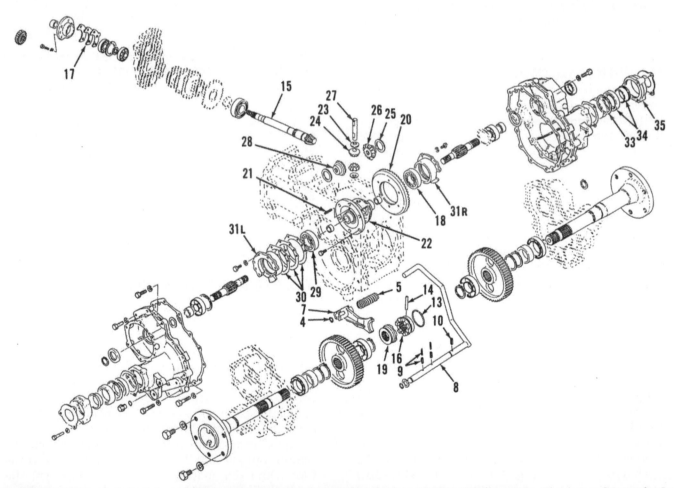

Fig. 90—Partially exploded view of differential, bevel pinion and final drive of 670 and 770 models. The two axles are locked and turn together when coupling (19) engages coupling (16).

4. Seal ring				26. Right side gear	
5. Spring	14. Pin	20. Ring gear		27. Cross shaft	
7. Differential lock fork	15. Bevel pinion shaft	21. Pin		28. Left side gear	
8. Differential lock shaft	16. Lock coupling	22. Differential housing		29. Ball bearing	
9. Roll pins	17. Shims	23. Thrust washer		30. Shims	
10. Roll pin	18. Ball bearing	24. Pinion		31L. Differential carrier housing (Left)	
13. Retainer ring	19. Locking collar	25. Thrust washer		31R. Differential carrier housing (Right)	

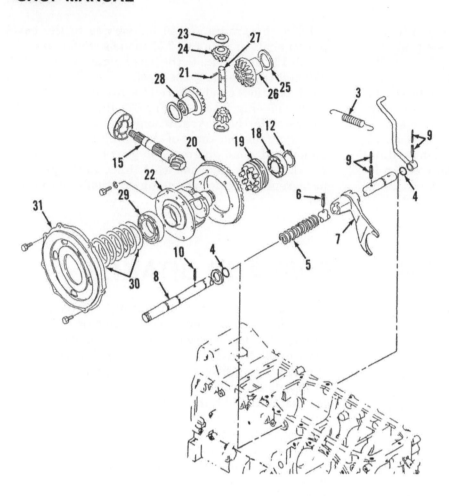

Fig. 91—Partially exploded view of differential, bevel pinion and final drive of 870, 970 and 1070 models. The differential is locked by sliding collar (19) through holes in housing (22) and into right side gear (26).

3. Spring	21. Pin
4. Seal ring	22. Differential
5. Spring	housing
6. Pin	23. Thrust washer
7. Differential lock fork	24. Pinion
8. Differential lock shaft	25. Thrust washer
9. Roll pins	26. Right side gear
10. Roll pin	27. Cross shaft
11. Washer	28. Left side gear
12. Snap ring	29. Ball bearing
15. Bevel pinion shaft	30. Shims
18. Ball bearing	31. Differential
19. Locking collar	carrier housing
20. Ring gear	

To remove the differential assembly, first drain oil from the transmission and hydraulic system, then remove rockshaft housing from top of center housing as outlined in paragraph 187 or paragraph 192. Remove axle assemblies as outlined in paragraph 159 or paragraph 162 and rear pto assembly from rear of center housing as outlined in paragraph 169 or paragraph 171. Unbolt and remove carrier housings (31), then lift differential assembly from center housing.

Ring gear (20) can be removed after removing the retaining screws. Drive pin (21) through housing bore and remove cross shaft (27), pinions (24), thrust washers (23), side gears (26 and 28) and thrust washers (25). Check all parts for wear, damage and freedom of movement.

Assemble differential using new spring pin (21). Clean threads in ring gear (20) and retaining screws with cleaner/primer and coat threads of retaining screws with medium strength thread lock. On 670 and 770 models, tighten ring gear retaining screws evenly to 54 N·m (40 ft.-lbs.) torque. On 870 model, tighten ring gear retaining screws evenly to 40 N·m (29 ft.-lbs.) torque. On 970 and 1070 models, tighten ring gear retaining screws evenly to 80 N·m (59 ft.-lbs.) torque. Make sure that ring gear is seated firmly and squarely against differential housing flange.

Shims (30) adjust backlash between ring gear (20) and bevel pinion (15). Install differential assembly (20 through 28) in center housing. Install bearing (29) and original or selected shims (30) in carrier housing. Coat edge of carrier housing (31), which pilots into bore of center housing, with grease before installing. Tighten screws retaining carrier housing evenly to 26 N·m (19 ft.-lbs.) torque.

To check backlash, push the installed differential assembly to left by tapping lightly against right side of differential housing (22) and bearing (18). Shims (30) are installed between left bearing carrier (31L) and rear axle center housing of 670 and 770 models. On 870, 970 and 1070 models, shims (30) are installed between bore of carrier housing (31) and bearing (29). Left side bearing (29) should be tight against shoulder of differential housing (22).

On all models, attach a dial indicator to measure movement at outside edge of ring gear teeth. Temporarily prevent bevel pinion from turning and measure backlash. Correct backlash is 0.13-0.18 mm (0.005-0.007 inch) for all models. Add or remove shims (30) as required to provide correct backlash. Shims are available in several thicknesses. Complete assembly by reversing disassembly.

158. Mesh position of the bevel pinion can be adjusted by adding or removing shims (17—Fig. 90) for 670 and 770 models or shims (10—Fig. 87) for 870, 970 and 1070 models. On all models, teeth of bevel pinion should contact center of ring gear teeth. To check mesh position of bevel pinion (15) with ring gear (20), paint some teeth of ring gear with machinist dye, then turn bevel pinion until teeth of bevel pinion pass the painted teeth. Observe mesh position as indicated by marks in paint. If mesh position is correct, marks from the bevel pinion teeth will be in the center of painted ring gear teeth.

Refer to paragraph 146 for service to the bevel pinion shaft of 670 and 770 models. Refer to paragraph 154 for service to the bevel pinion shaft of 870, 970 and 1070 range transmission. Bevel pinion (and mesh position) can be moved toward front by adding shims (17—Fig. 90 or 10—Fig. 87). Bevel pinion teeth will mesh farther to rear if some shims are removed. Changing the thickness of shims will move the bevel pinion and some difference in gear backlash will be noted. Recheck backlash as outlined in the previous paragraph 157 after changing position of bevel pinion.

REAR AXLE, FINAL DRIVE AND BRAKES

REAR AXLE AND FINAL DRIVE

670 And 770 Models

159. REMOVE AND REINSTALL. Final drive must be removed from right side before installing left final drive and axle assembly. Refer to paragraph 160.

To remove right side final drive and axle assembly, first lower the lift arms, then drain fluid from transmission and hydraulic system. Block front wheels to prevent rolling, then wedge between front axle and frame to prevent tipping. Raise tractor and support rear of tractor on jack stands, then remove rear wheel.

> **NOTE: Do not lift or support tractor under rear axle center housing. Weight of tractor may damage rear axle center housing if improperly lifted.**

Attach hoist to ROPS, if so equipped, then unbolt and remove the Roll Over Protection Structure from tractor. Unbolt and remove the 3-point hitch. Disconnect battery ground, remove rockshaft stop knob, rear pto knob and mid pto knob (if so equipped), then remove cover that surrounds these controls. Remove clutch housing cover and step plate, then unbolt brackets and fenders from step support or other brackets. Remove brake linkage and return spring from right pedal to right brake. Disconnect or remove wires, hydraulic lines and mid pto mounting brackets (if so equipped) that would interfere with removal of the right axle and final drive assembly. Make sure that rear of tractor is supported securely, then support final drive and axle assembly safely and in a way that will allow the right axle and final drive assembly to be separated from center housing. Remove attaching screws, then lift final drive and axle housing from center housing. Removal of left side final drive is similar to procedure for right side.

When reinstalling, clean mating surfaces and apply sealer, then tighten screws to 50 N·m (36 ft.-lbs.) torque. Tighten screws attaching drawbar bottom plate to 340 N·m (250 ft.-lbs.) torque. When attaching the "Roll-Gard" frame to models so equipped, be sure to install all screws, make sure that correct screws are installed and tighten screws evenly to the correct torque. On 670 and 770 models, tighten the 12 mounting screws to 265 N·m (195 ft.-lbs.) torque. The four screws attaching the upper crossbar should be tightened to 123 N·m (91 ft.-lbs.) torque. Tighten rear wheel retaining screws to 190 N·m (140 ft.-lbs.) torque.

160. The right final drive should be removed before attempting to install the left final drive. The sliding coupling (19—Fig. 90) of the differential lock is installed on left axle, and aligning groove of coupling with fork (7) is difficult unless right axle is removed. Refer to paragraph 159 for removal and installation procedure, but engage coupling groove with fork while positioning left final drive and axle on center housing.

161. OVERHAUL. To overhaul either left or right final drive and axle assembly, first remove the unit as outlined in paragraph 159 or paragraph 160. Remove retaining ring (13—Fig. 92), pin (14) and lock coupling (16) from right axle. Remove snap ring (33), bearing (34) and gear (35) from axle. Unbolt cover (45) and housing (44), then bump axle (32 or 53) from axle and final drive housing (39). Remove snap rings (36), then press sleeve (37), bearing (41), housing (44) and cover (45) from axle shaft. Seal (43) can be removed from housing (44) after removing snap ring (41). Inner sleeve of seal (43) can be pressed from axle. Open side of seal should be toward snap ring (42).

To remove pinion (52), first remove the axle as described, then refer to paragraph 165 for removal and service to the brake assemblies. Remove snap

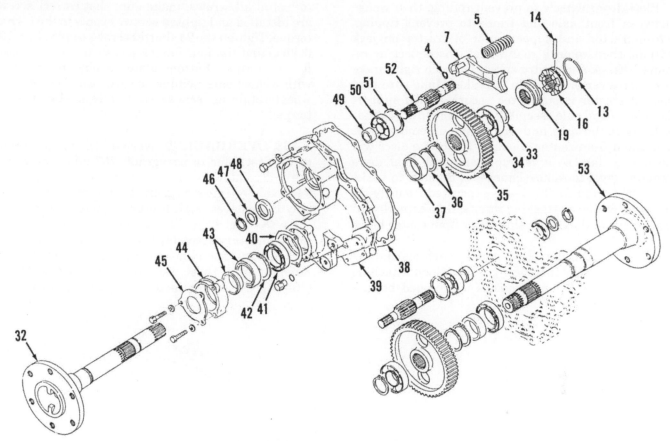

Fig. 92—Exploded view of typical 670 and 770 final drive and axle assemblies.

4. Seal ring	33. Snap ring	40. Gasket	47. Washer
5. Spring	34. Ball bearing	41. Ball bearing	48. Oil seal
7. Differential lock fork	35. Final drive	42. Snap ring	49. Sleeve
13. Retainer ring	reduction gear	43. Oil seal	50. Ball bearing
14. Pin	36. Snap rings	44. Housing	51. Snap ring
16. Lock coupling	37. Spacer	45. Cover	52. Final drive pinion
19. Locking collar	38. Gasket	46. Snap ring	53. Right axle
32. Left axle	39. Housing		

rings (46 and 51), then use a soft faced hammer to bump shaft and bearing (50) from housing (39). Press seal sleeve (49), bearing (50) and seal (48) from shaft or housing if renewal is required. Install new seal (48) with open side toward bearing (50) and coat lips of seal with grease before assembling.

Coat seal sleeve and seal lip with grease before assembling parts (42 - 45) on axle (32 or 53). Press bearing (41) against shoulder of axle. Install sleeve (37) and snap rings, then install axle assembly in housing (39) using a new gasket (40) coated with sealer. Tighten the three screws retaining cover (45) and housing (44) to 27 N·m (20 ft.-lbs.) torque. Heat bearing (34) to 150° C (300° F) in oil, then install gear (35), bearing and snap ring (33). If assembling right axle (53), align coupling (16) with hole in axle and coupling, then insert pin (14) and retaining ring (13).

Refer to paragraph 159 for installation of axle assembly. Brakes can be installed after installation of axle assemblies. The right final drive should re-

main removed until after the left final drive is installed. The differential lock sliding coupling (19) should be installed on left axle (32) while installing the final drive and axle assembly. Engage coupling groove with fork (7) while positioning left final drive and axle on center housing.

870, 970 And 1070 Models

162. REMOVE AND REINSTALL. To remove final drive and axle assembly from either side, first lower the lift arms, then drain fluid from transmission and hydraulic system. Unbolt and remove fenders and 3-point hitch. Attach hoist to ROPS, if so equipped, then unbolt and remove the Roll Over Protection Structure from tractor. If equipped with front-wheel drive, remove control rod and lower control lever. On all models, the hydraulic system selective control valve must be removed before removing the right axle housing.

Block front wheels to prevent rolling, then wedge between front axle and frame to prevent tipping. Raise tractor and support rear of tractor on jack stands, then remove rear wheel. Disconnect or remove linkage and return spring between right brake pedal or lever on left side to brake. Make sure that rear of tractor is supported securely, then support final drive and axle assembly safely and in a way that will allow the axle and final drive assembly to be separated from center housing. The brake discs will be free to fall, so it is important that final drive remains level to reduce chances of damage to brake parts when the final drive is separated from center housing. Remove attaching screws, then carefully lift final drive and axle housing away from center housing.

When reinstalling, clean mating surfaces and apply a flexible sealer to new gasket. When attaching the "Roll-Gard" frame to models so equipped, be sure to install all screws, make sure that correct screws are installed and tighten screws evenly to the correct torque. Tighten the 24 shorter screws to 136 N·m (100 ft.-lbs.) and the four longer screws to 332 N·m (245 ft.-lbs.) torque. Bottom plates should be installed with welded ears pointed toward rear. Tighten rear wheel retaining screws to 190 N·m (140 ft.-lbs.) torque.

163. OVERHAUL. To overhaul the removed final drive, first refer to paragraph 167 and remove the brake assembly. The input shaft (50—Fig. 93 or Fig. 94) is splined into side gears (25 and 28—Fig. 91) and may be removed with the brake assembly. Remove screw (8—Fig. 93 or Fig. 94) and washers, then withdraw planetary assembly from axle and ring gear (7).

Planetary assemblies for 870 and 970 models are different from those used on 1070 models. Shafts (13—Fig. 94) are retained by ring (24) on 870 and 970

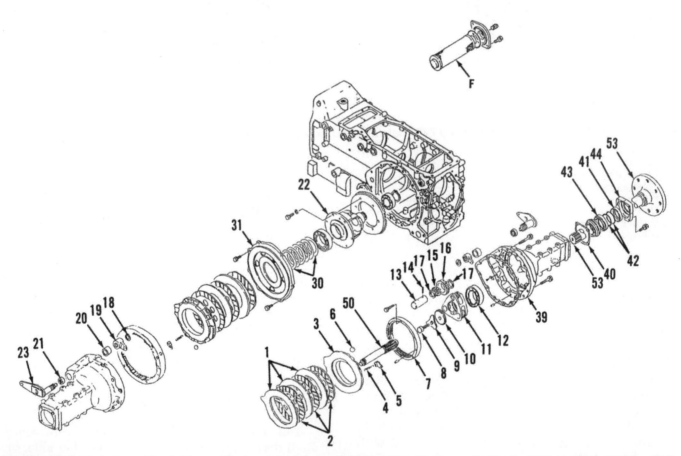

Fig. 93—Exploded view of final drive and axle assembly for 1070 models. Parts are similar for 870 and 970 models. Refer to Fig. 94 for planetary assembly typical of 870 and 970 models.

F. Suction filter	8. Screw	17. Thrust washer	31. Differential carrier housing
1. Brake plates (3 each side)	9. Washer	(2 for each gear)	39. Axle housing
2. Brake friction disks	10. Washer	18. Snap ring	40. Gasket
(3 each side)	11. Planet carrier	19. Brake cam	41. Dust seal
3. Brake actuator disk	12. Ball bearing	20. Bushing	42. "O" ring & packing ring
4. Return springs	13. Planet shaft	21. Oil seal	43. Oil seal
5. Spring clip	14. Roll pin	22. Differential housing	44. Housing
6. Actuator balls (6 each side)	15. Needle rollers	23. Brake actuator lever	50. Input shaft & sun gear
7. Gear	16. Planet gear	30. Shims	53. Right axle

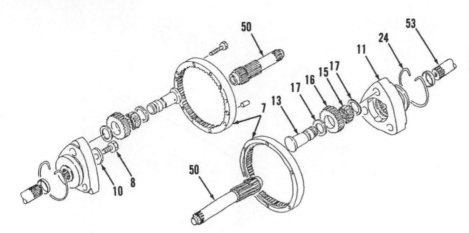

Fig. 94—Exploded view of planetary reduction typical of 870 and 970 models.

8. Screw
10. Washer
11. Planet carrier
13. Planet shaft
15. Needle rollers
16. Planet gear
17. Thrust washer (2 for each gear)
24. Retaining ring
50. Input shaft & sun gear
53. Right axle

models. Shafts (13—Fig. 93) are retained by pins (14) on 1070 models. Remove retaining ring (24—Fig. 94) or pin (14—Fig. 93), then press shafts (13—Fig. 93 or Fig. 94) from planet carrier (11). Needle bearings (15) contain 30 loose rollers on 1070 models.

Lubricate all parts with clean transmission oil and reassemble by reversing disassembly procedure. Make sure that retaining ring (24—Fig. 94) is correctly seated.

BRAKES

670 And 770 Models

164. ADJUSTMENT. Each brake pedal should have 35 mm (1⅜ in.) travel and pedals should align when equal pressure is applied to the two pedals. Adjust by shortening or lengthening actuating rods. Loosen locknuts (1—Fig. 95) and turn the linkage turnbuckles (2 and 3) until pedal travel is correct. Tighten locknuts to maintain adjustment and to align end yokes (4).

165. R&R AND OVERHAUL. Remove the rear wheel, detach yoke from cam lever (1—Fig. 96), then unbolt and remove cover (3). Remove snap ring (8), then pull drum (9) from pinion shaft splines. A washer is located on pinion shaft, behind drum. Remainder of disassembly will depend upon extent of repair, but will be evident.

Use a new gasket (4) when assembling. Tighten wheel retaining screws to 190 N·m (140 ft.-lbs.) torque and adjust brakes as outlined in paragraph 164.

870, 970 And 1070 Models

166. ADJUSTMENT AND LINKAGE. Each brake pedal should have 35 mm (1⅜ in.) travel and pedals should align when equal pressure is applied to the two pedals. Adjust by shortening or lengthening actuating rods. Loosen locknuts (1—Fig. 95) and turn

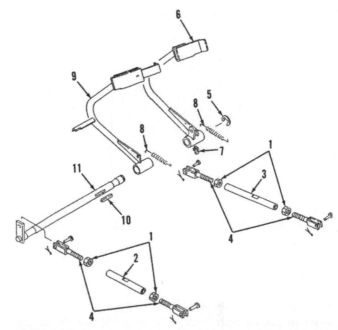

Fig. 95—Brake linkage for all models is similar. Adjustment is accomplished by changing length of connecting linkage.

1. Locknuts
2. Turnbuckle
3. Turnbuckle
4. Yokes
5. "E" ring
6. Right brake pedal
7. Grease fitting
8. Return springs
9. Left brake pedal
10. Key
11. Left brake cross shaft

the linkage turnbuckles (2 and 3) until pedal travel is correct. Tighten locknuts to maintain adjustment and to align end yokes (4).

Left side bushing for cross shaft (11) should be installed 25 mm (0.984 in.) deep and right side bushing should be installed to depth of 32 mm (1.250 in.). Right pedal (6) should move freely on end of cross shaft and should be greased regularly through fitting (7).

167. R&R AND OVERHAUL. Brake discs (2—Fig. 93) and separator plates (1) can be renewed after

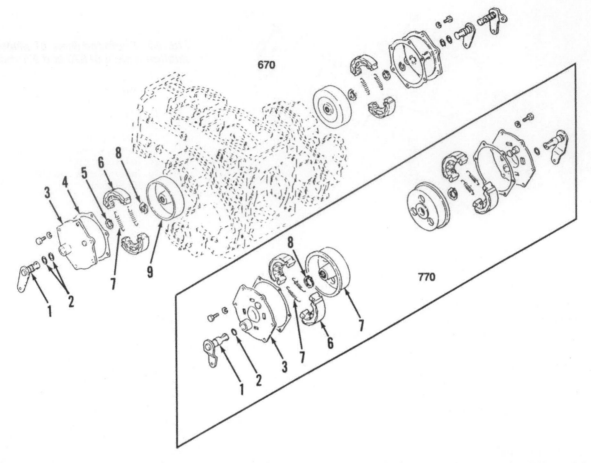

Fig. 96—Exploded view of drum type brake assemblies for 670 model. Inset shows similar parts for 770 model.

1. Cam lever
2. "O" rings
3. Cover
4. Gasket
5. Snap ring
6. Brake shoe
7. Spring
8. Snap ring
9. Brake drum

removing rear axle and final drive housing as outlined in paragraph 162. Standing the removed axle and final drive on the wheel flange may make disassembly and reassembly easier.

If not removed with discs (2) and separator plates (1), lift pinion shaft (50) from final drive. Disconnect springs (4) and lift actuator plate (3) from housing. Balls (6) will fall free. **Be sure to locate all six balls before assembling.** If necessary, cam (19), bushing (20) and seal (21) can be removed for service after removing snap ring (18). Lip of seal (21) should be

coated with multipurpose grease and should be installed with lip toward center housing. Use petroleum jelly to stick balls (6) in position while assembling. Cam (19) and shaft (23) have marks which should be matched when assembling. Install pinion shaft (50), friction disks (2) and separator plates (1), alternating discs and plates. Make sure that lug on separator plates correctly engages cut-out in final drive housing. Refer to paragraph 162 for installation of axle and final drive housing. Adjust brake linkage as outlined in paragraph 166.

POWER TAKE-OFF

168. The pto system consists of the mechanical components necessary to drive both mid and rear output shafts, components to control engagement, and the necessary mechanical and electrical components required to safely control operation. Continuous live pto is provided on some models by a dual stage engine clutch. Control of pto is by engaging or disengaging a separate jaw type pto clutch. Refer to paragraph 135 and following for service to engine

clutches and to paragraph 139 for adjusting the dual stage clutch of models so equipped.

REAR PTO

670 And 770 Models

169. R&R AND OVERHAUL. To remove the rear pto shift linkage, remove rockshaft stop knob, rear

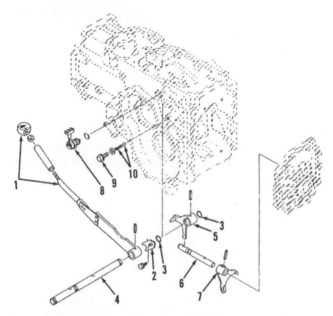

Fig. 97—Exploded view of rear pto engagement linkage, rail and fork for 670 and 770 models.

1. Engagement lever	6. Rail
2. Retainer plate	7. Fork
3. "O" rings	8. Safety switch
4. Cross shaft	9. Detent plug
5. Change arm	10. Ball & spring

pto knob and mid pto knob (if so equipped), then remove cover that surrounds these controls. Raise and support rear of tractor safely, then remove both rear wheels. Unbolt and remove fender from left side. Engagement lever (1—Fig. 97) can be pulled from shaft after removing roll pin. Safety switch (8) should have continuity between terminals only when pto is in disengaged position and should not be grounded. Switch should prevent starting unless pto is disengaged.

To service parts (3, 4 and 5), remove engagement lever (1), safety switch (8), plug (9) and detent (10), then refer to paragraph 187 and remove the rockshaft lift cover. It may be necessary to use a magnet to remove the detent ball from bore in housing. Drive roll pin from arm (5), remove retainer (2), then drive cross shaft toward right side of tractor until removed from housing and arm. When assembling, insert cross shaft from right side.

To service shift rail (6) and fork (7), first remove the cross shaft (4) and arm (5) as previously described, then refer to paragraph 170 and remove the rear pto driveshaft assembly.

170. The pto system is driven by shaft (1—Fig. 98 or Fig. 99) from the engine clutch. Engagement linkage is similar for all 670 and 770 models. Some tractors are equipped with a dual stage engine clutch which provides live pto and parts are shown in Fig. 99. Tractors equipped with a single stage engine clutch use pto parts shown in Fig. 98.

To remove the rear pto shaft and related parts, first remove the rockshaft lift cover as outlined in paragraph 187 and the pto engagement linkage as outlined in paragraph 169. Unbolt and remove the 3-point hitch assembly and the shield from around the rear pto shaft. Unbolt the drawbar plate and mid-pto covers (if so equipped), then unbolt and remove the rear cover (9—Fig. 98 or Fig. 99). Parts (6 through 30) will be removed with rear cover, along with rail (6—Fig. 97) and fork (7). Remove drive pinion and shaft (7—Fig. 98 or Fig. 99), then press output shaft (30) and associated parts from rear cover (9).

When assembling, reverse disassembly procedure. Be sure to coat parts with oil when assembling. Clean mating surfaces and coat new gasket with nonhardening sealer before installing the rear cover. Coat threads of all screws retaining rear cover (9) with medium strength thread lock, then tighten the six larger screws to 265 N·m (195 ft.-lbs.) torque. Install and tighten others screws evenly. Coat threads of drawbar plate retaining screws with medium strength thread lock, then tighten to 340 N·m (250 ft.-lbs.) torque. Refer to paragraph 169 and Fig. 97 when installing the engagement linkage.

870, 970 And 1070 Models

171. R&R AND OVERHAUL. The pto hand lever (1—Fig. 100) and link (9) can be removed after removing rockshaft stop knob, rear pto knob and mid pto knob (if so equipped), then removing the cover that surrounds these controls.

To service parts (3, 4 and 5), first remove rockshaft stop knob, rear pto knob and mid pto knob (if so equipped), then remove the cover that surrounds these controls. Disconnect link (9), then refer to paragraph 192 and remove the rockshaft lift cover. Remove roll pin, then pull lever (4) from shaft (5). Push shaft (5) in and remove.

To service shift rail (6) and fork (7), first remove lever (4) and shaft (5) as previously described, then refer to paragraph 172 and remove the rear pto driveshaft assembly.

172. The pto system is driven by shaft (1—Fig. 101) from the dual stage engine clutch which provides live pto. To remove the rear pto shaft and related parts, first remove the rockshaft lift cover as outlined in paragraph 192 and the pto engagement linkage as outlined in paragraph 171. Unbolt and remove the 3-point hitch assembly, the drawbar assembly and the shield from around the rear pto shaft. If equipped with mid-pto, remove the shaft assembly. On all models, unbolt and remove the rear cover (9), with parts (6 through 30). If front coupling (5) falls into transmission housing, lower cover can be removed to replace the coupling before installing. Remove drive

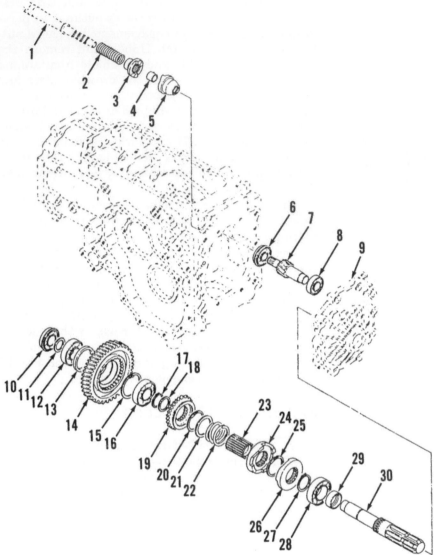

Fig. 98—Exploded view of rear pto drive for 670 and 770 models with single stage engine clutch.

1. Drive shaft	18. Guide
2. Spring	19. Engagement
3. Drive coupling	coupling
4. Bushing	20. Snap ring
5. Coupling	(same as 25)
6. Ball bearing	21. Spacer
7. Drive shaft & gear	22. Spring
8. Ball bearing	23. Slider sleeve
9. Rear cover	24. Overrunning clutch
10. Ball bearing	front half
11. Spacer	25. Snap ring
12. Ball bearing	(same as 20)
(same as 16)	26. Overrunning clutch
13. Spacer	rear half
14. Driven gear	27. Snap ring
15. Snap ring	28. Ball bearing
16. Ball bearing	29. Sleeve
(same as 12)	30. Rear pto
17. Snap ring	output shaft

pinion and shaft (7), then press output shaft (30) and associated parts from rear cover (9).

When assembling, reverse disassembly procedure. Be sure to coat parts with oil when assembling. Clean mating surfaces and coat new gasket with nonhardening sealer before installing the rear cover. Coat threads of all screws retaining rear cover (9) with medium strength thread lock, then tighten the six larger screws to 265 N·m (195 ft.-lbs.) torque. Install and tighten others screws evenly. Coat threads of drawbar plate retaining screws with medium strength thread lock, then tighten to 340 N·m (250 ft.-lbs.) torque. Refer to paragraph 171 and Fig. 100 when installing the engagement linkage.

MID PTO

670 And 770 Models

173. CONTROLS. Refer to Fig. 102 for drawing of mid-pto clutch and engagement linkage. To remove or service the mid-pto shift linkage, remove the lower dash and clutch housing covers. Raise and support rear of tractor safely, then remove the left rear wheel, left fender and drawbar assembly. Disconnect sway links, remove the lower draft arm and remove gearbox covers. Loosen spring retainer (2—Fig. 103) until spring (3) is free, then remove linkage as required for inspection or service.

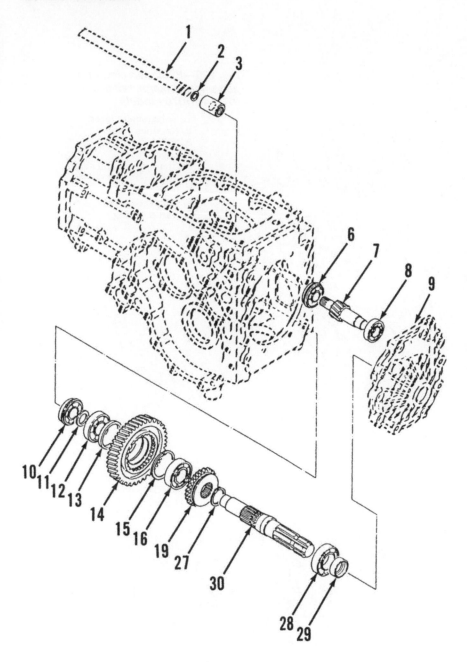

Fig. 99—Exploded view of pto drive for 670 and 770 models with live pto. A dual stage engine clutch is used.

1. Drive shaft
2. Snap ring
3. Drive coupling
6. Ball bearing
7. Drive shaft & gear
8. Ball bearing
9. Rear cover
10. Ball bearing
11. Spacer
12. Ball bearing
   (same as 16)
13. Spacer

14. Driven gear
15. Snap ring
16. Ball bearing
   (same as 12)
19. Engagement
   coupling
27. Snap ring
28. Ball bearing
29. Sleeve
30. Rear pto
   output shaft

Adjust length of spring (3) by turning retainer (2). Correct length of spring is 59 mm (2⅜ in.) when lever (1) is down in the disengaged position. Adjust length of rod (4) so that control lever (1) does not contact top of slot when engaged or bottom of slot when disengaged. Turn nut (5) to adjust length of the spring capsule (6) to just fit between bellcranks (8 and 10) when mid-pto lever and engine clutch pedal are both engaged.

174. R&R AND OVERHAUL MID-PTO GEARBOX. To remove the mid-pto (Fig. 104), drain oil from the transmission housing and remove drawbar assembly. Disconnect left side sway link, then unbolt

and remove the lower draft link and bracket from left side. Remove wiring cover (34) from mid-pto gearbox, then disconnect wiring from switch (31). Detach control linkage rods (4 and 9—Fig. 103), then unbolt and remove the gearbox (Fig. 104).

Remove pin (23), then use a brass drift to remove shaft (24). Remove idler gear (27), needle bearings (26) and washers (24 and 25). Unbolt lower cover (32) from gearbox and remove pins (11 and 12—Fig. 103). Unbolt and remove retainer plate (16), then drive the brake operating shaft (15) from brake lever (13). Remove pins (18 and 19), then unbolt and remove retainer plate (24). Drive the engagement shaft (23) from shift arm (20).

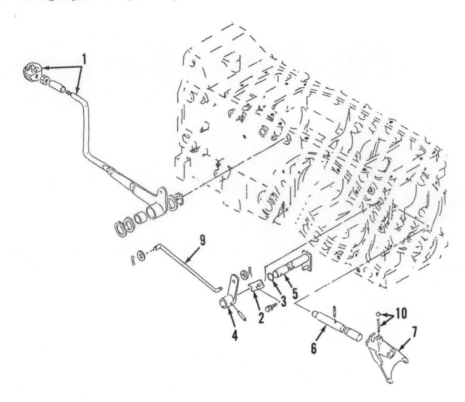

Fig. 100—Exploded view of rear pto engagement linkage, rail and fork for 870, 970 and 1070 models.

1. Engagement lever
2. Retainer plate
3. "O" ring
4. Lever
5. Change arm and shaft
6. Rail
7. Fork
9. Connecting link
10. Ball & spring

Fig. 101—Exploded view of pto drive for 870, 970 and 1070 models. A dual stage engine clutch is used which provides live pto.

1. Drive shaft
2. Snap ring
3. Drive coupling
4. Snap ring
5. Coupling
6. Ball bearing
7. Drive shaft & gear
8. Ball bearing
9. Rear cover
10. Ball bearing
11. Spacers
12. Ball bearing (same as 16)
14. Driven gear
15. Snap rings
16. Ball bearing (same as 12)
27. Snap rings
28. Ball bearing
29. Seal
30. Rear pto output shaft

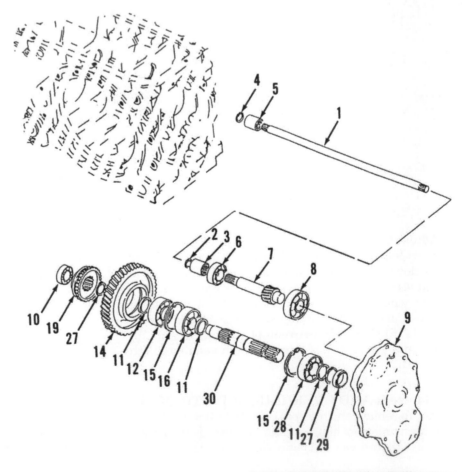

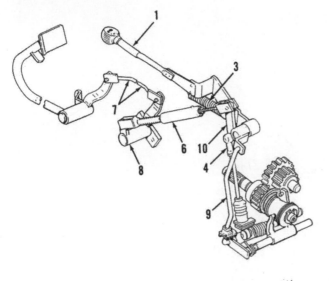

Fig. 102—Drawing of mid-pto control linkage for 670 and 770 models so equipped. Refer to Fig. 103 for legend.

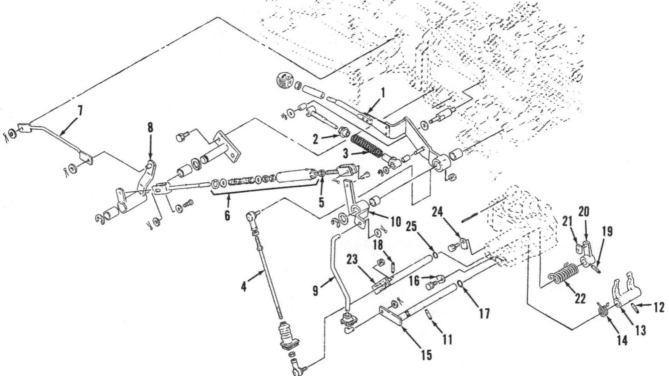

Fig. 103—Exploded view of control linkage for mid-pto available on 670 and 770 models.

1. Engagement lever
2. Spring retainer
3. Spring
4. Pto clutch rod
5. Adjusting nut
6. Spring capsule
7. Engine clutch rod
8. Bellcrank
9. Mid-pto brake rod
10. Bellcrank
11. Pin
12. Pin
13. Brake lever
14. Return spring
15. Brake operating shaft
16. Retainer plate
17. "O" ring
18. Pin
19. Pin
20. Shift arm
21. Shifter
22. Spring
23. Engagement shaft
24. Retainer plate
25. "O" ring

Remove covers (1 and 2—Fig. 104), then withdraw output shaft (14) and related parts. Be careful not to lose the spring loaded detent balls (16) when disassembling. Inspect for worn or damaged parts and renew as necessary.

Assemble gearbox by reversing the disassembly procedure. Lubricate lip of seal (3) with grease. Use new gasket (33) and "O" ring (30) when installing the mid-pto gearbox. Tighten gearbox retaining screws to 50 N·m (37 ft.-lbs.) torque and the drawbar retaining screws to 340 N·m (250 ft.-lbs.) torque. Refer to paragraph 139 for adjusting the dual stage clutch (of models so equipped) and to paragraph 173 for adjusting pto control linkage.

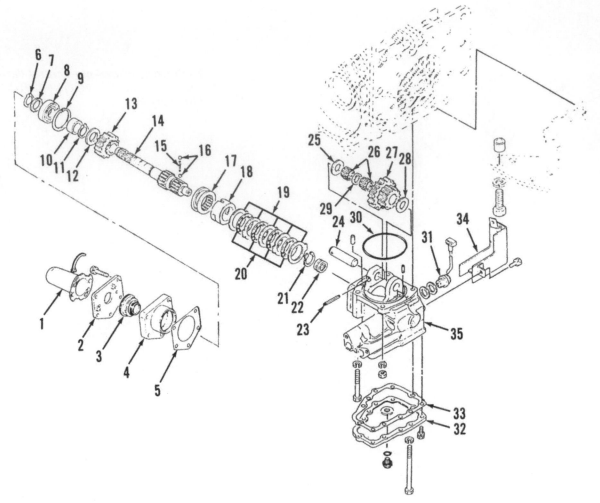

Fig. 104—Exploded view of mid-pto drive housing and gears available on 670 and 770 models.

1. Cover
2. Cover
3. Oil seal
4. Seal case
5. Gasket
6. Snap ring
7. Spacer
8. Ball bearing
9. Snap ring
10. Bushing
11. Snap ring
12. Washer
13. Driven gear
14. Output shaft
15. Detent spring
16. Detent balls
17. Sliding collar
18. Actuator
19. Plates (5 used)
20. Brake discs (4 used)
21. Snap ring
22. Needle bearing
23. Pin
24. Shaft
25. Washer (same as 28)
26. Needle bearings
27. Idler gear
28. Washer (same as 25)
29. Spacer
30. "O" ring
31. Switch
32. Lower cover
33. Gasket
34. Cover
35. Housing

870, 970 And 1070 Models

175. CONTROLS. To remove or service the mid-pto shift linkage, remove the left draft arm and link assembly from the 3-point hitch. Raise and support rear of tractor safely, then remove the left rear wheel, left fender, tool box and left side step plate. Remove gearbox cover, loosen spring retainer (2—Fig. 105) until spring (3) is free, then remove linkage as required for inspection or service.

Adjust length of spring (3) by turning retainer (2). Correct length of spring is 59 mm (2⅜ in.) when lever (1) is down in the disengaged position. Adjust length of rod (4) so that control lever (1) does not contact top of slot when engaged or bottom of slot when disengaged. Turn nut (5) to adjust length of the spring

capsule (6) so that 15 mm (⁹/₁₆ in.) gap exists at front of slot in rod (7) when mid-pto lever and engine clutch pedal are both engaged.

176. R&R AND OVERHAUL MID-PTO GEAR-BOX. To remove the mid-pto (Fig. 106), first drain oil from the transmission housing. Unbolt and remove cover (1), then remove idler shaft (24). Remove wiring cover (34—Fig. 105), disconnect wiring from switch (31—Fig. 106), then pull wire and connector free. Detach control linkage rods (4 and 9—Fig. 105) from levers (44 and 50—Fig. 106), then unbolt and remove the gearbox.

Unbolt and remove retainer plate (43), then drive pins (36, 37 and 38) from engagement shaft (44) and levers (39 and 41). Drive engagement shaft (44) from

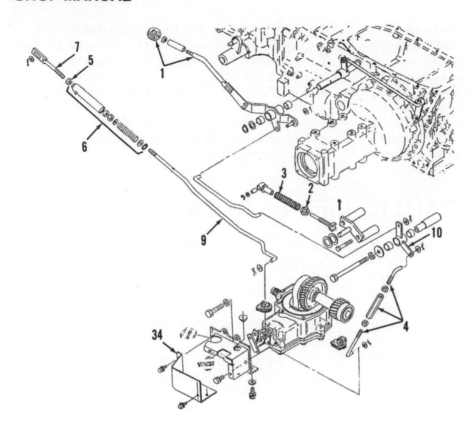

Fig. 105—Exploded view of control linkage for mid-pto available on 870, 970 and 1070 models.

1. Engagement lever
2. Spring retainer
3. Spring
4. Pto clutch rod
5. Adjusting nut
6. Spring capsule
7. Engine clutch rod
9. Mid-pto brake rod
10. Bellcrank
34. Cover

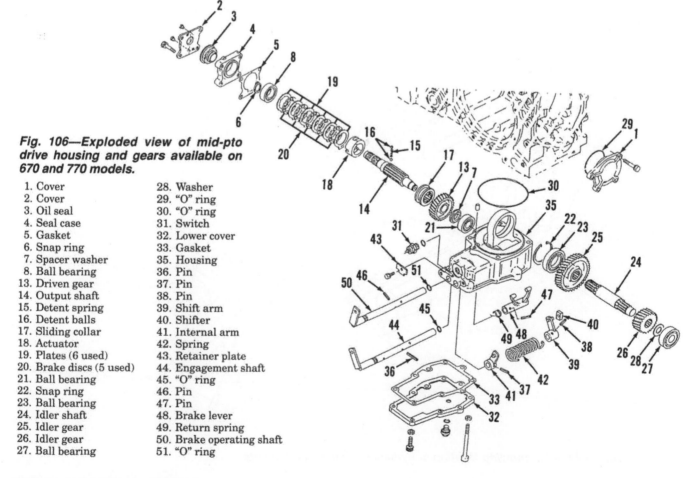

Fig. 106—Exploded view of mid-pto drive housing and gears available on 670 and 770 models.

1. Cover	28. Washer
2. Cover	29. "O" ring
3. Oil seal	30. "O" ring
4. Seal case	31. Switch
5. Gasket	32. Lower cover
6. Snap ring	33. Gasket
7. Spacer washer	35. Housing
8. Ball bearing	36. Pin
13. Driven gear	37. Pin
14. Output shaft	38. Pin
15. Detent spring	39. Shift arm
16. Detent balls	40. Shifter
17. Sliding collar	41. Internal arm
18. Actuator	42. Spring
19. Plates (6 used)	43. Retainer plate
20. Brake discs (5 used)	44. Engagement shaft
21. Ball bearing	45. "O" ring
22. Snap ring	46. Pin
23. Ball bearing	47. Pin
24. Idler shaft	48. Brake lever
25. Idler gear	49. Return spring
26. Idler gear	50. Brake operating shaft
27. Ball bearing	51. "O" ring

shift arms (39 and 41) and housing. Drive pins (46 and 47) from the brake operating shaft (50) and brake lever (48), then drive shaft from lever and housing.

Remove cover (2), then withdraw output shaft (14) and related parts. Be careful not to lose the spring loaded detent balls (16) when disassembling. Inspect all parts for wear or damage and renew as necessary.

Assemble gearbox by reversing the disassembly procedure. Install sliding collar (17) with bevel side toward driven gear (13). Use new gasket (33) and "O" ring (30) when installing the mid-pto gearbox. Tighten socket head screws retaining gearbox to 75 N·m (55 ft.-lbs.) torque, hex head retaining screws to 50 N·m (37 ft.-lbs.) torque and the rear cover retaining screws to 27 N·m (20 ft.-lbs.) torque. Refer to paragraph 139 for adjusting the dual stage clutch (of models so equipped) and to paragraph 175 for adjusting pto control linkage.

HYDRAULIC LIFT SYSTEM

177. The hydraulic lift system basically consists of an oil reservoir (or sump), hydraulic pump, control valve, lift cylinder and lift links (Fig. 107). The hydraulic lift pump is mounted on left side of engine and is driven by the camshaft gear whenever the engine is running. The pump supplies pressurized fluid for rockshaft control, remote control valves. Models with power steering are equipped with a second pump mounted ahead of the lift system pump.

FLUID AND FILTERS

All Models

178. Oil used as operating fluid for the hydraulic lift system also serves as the lubricant for the trans-mission, final drive and rear axles. The transmission housing, rear axle center housing and axle housings are used as the reservoir to contain the hydraulic system operating fluid. Oil level should be maintained between the marks on the dipstick. Refer to Fig. 108A, Fig. 108B and Fig. 109 for location of the dipstick and fill plug.

The manufacturer recommends changing fluids and servicing filters after each 500 hours of normal operation. The suction screen can be cleaned, but new cartridge filter should be installed. Service filters and change hydraulic fluid more frequently if operating in dusty conditions. Lower the lift arms before draining oil. Refer to paragraph 36 for draining oil and servicing suction screen. Oil is drained by removing plugs located in the bottom of transmission and rear axle center housing. Some models are equipped with

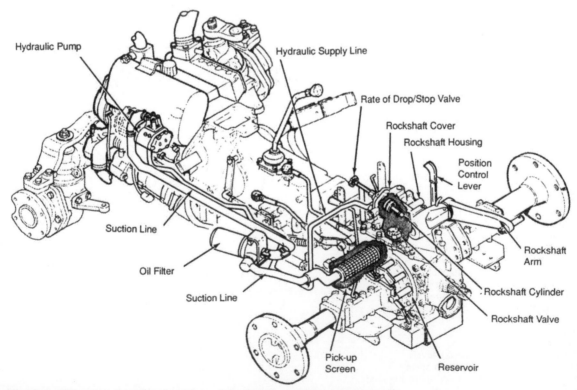

Fig. 107—Drawing of tractor showing location of hydraulic system components.

Fig. 108A—View of dipstick (D) for transmission and hydraulic fluid typical of 670 and 770 models.

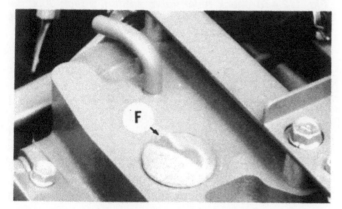

Fig. 108B—Oil filler cap (F) is located at top rear corner of rockshaft housing on 670 and 770 models.

Fig. 109—View of dipstick (D) and filler plug (F) for transmission and hydraulic fluid typical of 870, 970 and 1070 models.

a canister type filter located in the hydrostatic steering circuit, but all models are equipped with suction screen (filter). All filters should be serviced when oil is drained and reservoir is filled with new oil.

Fill system reservoir with "John Deere Low Viscosity HY-GARD" or equivalent fluid, then start engine. Some oil may remain in system when draining, so it is important to check level with dipstick when filling. Allow engine to run at idle while operating hydraulic

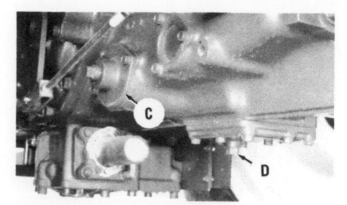

Fig. 110—View of cover (C) which must be removed to clean the suction screen.

controls and raising and lowering lift arms. Stop engine, lower three-point hitch, allow sufficient time for oil level to stabilize between compartments and check fluid level. Tractor should be on level surface when checking fluid level.

TESTS AND ADJUSTMENTS

All Models

179. Before conducting any pressure or flow checks, make sure that the correct type and amount of fluid is contained in the reservoir. Refer to paragraph 178 for servicing fluids and filters. Check for external oil leaks. Inspect for bent, disconnected or improperly adjusted linkage and for dirt or other objects that would prevent normal movement of linkage. Tests should not be considered accurate until fluid reaches normal temperature. Oil that is too hot or too cool may result in inaccurate pressures and flow.

Models Without Selective Control Valve

180. FLOW AND RELIEF PRESSURE TESTS. To check pump flow, attach adapter to bottom (outlet) of pump and connect to flow meter inlet. Return fluid from flow meter can be directed back to reservoir through filler plug opening (F—Fig. 108B or Fig. 109). Operate engine at 2600 rpm and slowly close flow meter restriction until pressure is increased to 10,340 kPa (1500 psi). Pumps on 670 and 770 models should have at least 17 LPM (4.5 GPM). Pumps on 870, 970 and 1070 models should provide minimum of 26.5 LPM (7.0 GPM). If flow is low, check and clean inlet suction screen (Fig. 110). Air leaks or restrictions in the suction line can also reduce pump output.

To check main relief pressure, connect 20,000 kPa (3000 psi) pressure gauge as shown in Fig. 111 or Fig. 112. Lower the rockshaft, run engine at high idle speed and close the drop/stop valve by turning knob (R) clockwise as far as possible. Pull rockshaft posi-

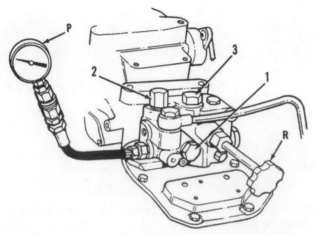

Fig. 111—To check relief valve setting, attach 3000 psi pressure gauge (P) as shown to 670 and 770 models without selective control valve.

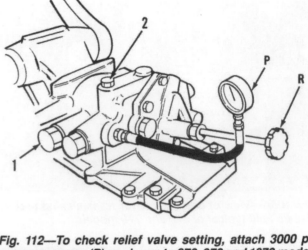

Fig. 112—To check relief valve setting, attach 3000 psi pressure gauge (P) as shown to 870, 970 and 1070 models without selective control valve.

tion lever back until system relief valve opens and observe pressure on gauge. System relief pressure should be 13,790-14,480 kPa (2000-2100 psi) for all 670 and 770 models. System relief pressure should be 15,170-15,860 kPa (2200-2300 psi) for all 870, 970 and 1070 models. If pressure is too low, install plug at port (1—Fig. 111 or Fig. 112) and recheck. If pressure is still too low, the pressure relief valve may be set too low or valve may be leaking. Refer to paragraph 184 for changing relief setting. If pressure increases when port (1) is plugged, leakage of the rockshaft valve is indicated.

Models With Selective Control Valve

181. FLOW AND RELIEF PRESSURE TESTS. Pump flow can be checked by attaching flow meter to front connector outlet (O—Fig. 113). Return fluid

from flow meter can be directed back to reservoir through filler plug opening (F). Operate engine at 2600 rpm and slowly close flow meter restriction until pressure is increased to 10,340 kPa (1500 psi). Pumps on 670 and 770 models should have at least 17 LPM (4.5 GPM). Pumps on 870, 970 and 1070 models should provide minimum of 26.5 LPM (7.0 GPM). If flow is low, check and clean inlet suction screen (Fig. 109). Air leaks or restrictions in the suction line can also reduce pump output.

To check main relief pressure on models with selective control valve, attach flow meter to front connector outlet as shown in Fig. 113 and close valve. Run engine at high idle speed and operate the selective control valve to pressurize the outlet with test gauge (P). Observe pressure indicated on gauge. System relief pressure should be 13,790-14,480 kPa (2000-2100 psi) for all 670 and 770 models. System

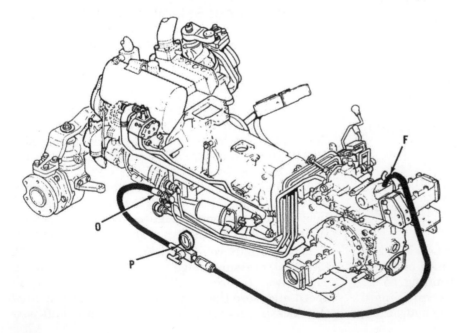

Fig. 113—View of flow meter or pressure gauge (P) connected to front connector outlet (O) models with selective control valve. Fluid can be directed back to filler plug opening (F).

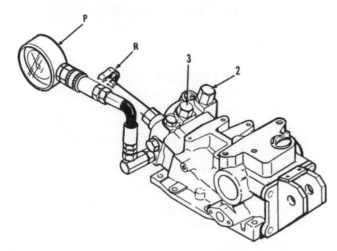

Fig. 114—View showing gauge connection for checking 670 and 770 models for rockshaft leakage.

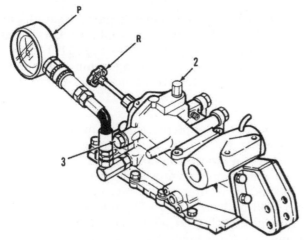

Fig. 115—View showing gauge connection for checking 870, 970 and 1070 models for rockshaft leakage.

relief pressure should be 15,170-15,860 kPa (2200-2300 psi) for all 870, 970 and 1070 models. If pressure is low, check for rockshaft leakage. Relief setting can be adjusted as outlined in paragraph 184.

Refer to Fig. 114 or Fig. 115 and attach a 20,000 kPa (3000 psi) pressure gauge (P) as shown. Lower the rockshaft, run engine at high idle speed and close the drop/stop valve by turning knob (R) clockwise as far as possible. Pull rockshaft position lever back until system relief valve opens, and observe pressure. If pressure is not 13,790-14,480 kPa (2000-2100 psi) for 670 and 770 models or 15,170-15,860 kPa (2200-2300 psi) for 870, 970 and 1070 models, remove plug (3) and add or remove shims located between spring and plug. Recheck pressure after changing thickness of shims. If pressure is within limits, observe pressure indicated on gauge and shut engine off. Pressure should not drop faster than 700 kPa (100 psi) in 1 minute. If pressure drops too fast, check for leaking rockshaft relief valve under plug (3) and for leaking

seal between rockshaft valve and housing. Leakage may also be internal, around lowering valve or load check valve.

Open the rate of drop/stop valve by turning knob (R) counter-clockwise as far as possible, run engine at high idle speed, adjust the rockshaft position sensing feedback linkage until system is in relief indicated by relief valve opening, then observe pressure on gauge and shut engine off. Pressure should not drop faster than 1400 kPa (200 psi) in 1 minute. If pressure drops too fast, there may be leakage past the rockshaft relief valve located under plug (3) or past the piston in the rockshaft cylinder.

All Models

182. ROCKSHAFT LEAKAGE TEST. A 227 kg (500 lbs.) weight should be attached to the 3-point lift arms to check for system leakage. Open the drop/rate valve completely by turning knob (R—Fig. 114 or Fig. 115) counterclockwise as far as possible, run the engine at high idle speed, pull the position control lever back and observe the time required to fully raise the weight. Arms should raise from completely lowered to fully raised in 2½-3 seconds. If other tests are correct, but the time to raise the weight is too long, the rockshaft valve is probably leaking and should be serviced. If the lift arms do not raise completely before the system relief valve opens, check adjustment of position sensing linkage.

183. POSITION CONTROL ADJUSTMENT. Start engine and operate at slow idle speed, then move position control lever fully to the rear to raise the lift arms. Loosen locknuts on lower end of link (L—Fig. 116), then turn nuts to lengthen link until relief valve opens. Shorten link by turning upper nut two turns, then tighten nuts to maintain adjustment. When link is correctly adjusted, lift arms should have 3 mm (⅛ inch) free play at top of travel.

184. PRESSURE RELIEF SETTING. Refer to paragraph 180 or paragraph 181 for checking relief pressure. System relief pressure should be 13,790-14,480 kPa (2000-2100 psi) for all 670 and 770 models. System relief pressure should be 15,170-15,860 kPa (2200-2300 psi) for all 870, 970 and 1070 models. If pressure is too low, remove plug (2—Fig. 114 or Fig. 115), add shims between plug and spring, then reinstall plug. Recheck relief pressure after adding or removing shims.

1070 Models

185. DRAFT CONTROL ADJUSTMENT. Draft sensing feedback link (D—Fig. 116) should be adjusted to provide full range of sensitivity without causing the system to open the pressure relief valve.

To adjust, start engine and operate at slow idle speed, then move position control lever fully to the rear to raise the lift arms. Loosen locknut at rear of link (D), then lengthen link until relief valve just opens. Shorten link 3½ turns and tighten locknut.

Fig. 116—Adjust length of position control link (L) by turning nuts on lower end as described in text.

HYDRAULIC PUMP

All Models

186. R&R AND OVERHAUL. To remove the hydraulic pump from left side of engine, first remove engine shield from left side of engine, then detach suction tube from top of pump and outlet line from underside of pump. Unbolt and remove flange from top of pump, then unbolt and remove hydraulic pump from timing gear housing. Inspect splines of pump driveshaft and coupling in drive gear.

Scribe a mark across pump housing and cover to ensure correct reassembly. Unbolt and remove pump cover (1—Fig. 117 or Fig. 118). Remove seals (3) and gears (5 and 6). Remove snap ring (10) and driveshaft seal (9). Make sure that all parts, especially the small fluid passages are clean, then inspect all parts for scoring, wear or damage.

Renew all seals and damaged parts. Coat all parts with hydraulic fluid when assembling. On 670 and 770 models, bearing plates (4—Fig. 117) must be installed with brass side facing pump gears and oil groove positioned toward inlet side of pump. Press pump driveshaft seal (9—Fig. 117 or Fig. 118) into bore until just below snap ring groove, then install snap ring (10). Complete assembly making sure that all packing rings and back-up rings are properly positioned in groove. Install cover (1 or 11), aligning scribe marks on housing and cover made prior to disassembly. Tighten cover retaining screws to 15 N•m (133 in.-lbs.) torque. Tighten screws attaching pump inlet fitting to pump to 6 N•m (53 in.-lbs.) torque.

When installing pump, tighten pump attaching screws and nuts to 26 N•m (230 in.-lbs.) torque.

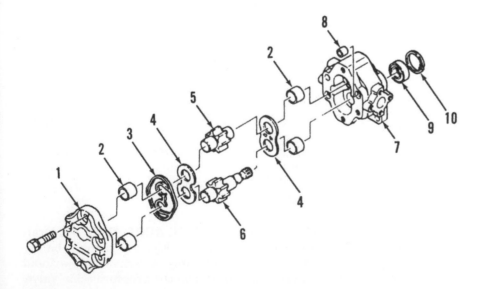

Fig. 117—Exploded view of lift system pump used on 670 and 770 models.

1. Cover
2. Bushings
3. Packing
4. Bearing plates
5. Driven gear
6. Drive gear & shaft
7. Pump body
8. Dowel
9. Oil seal
10. Snap ring

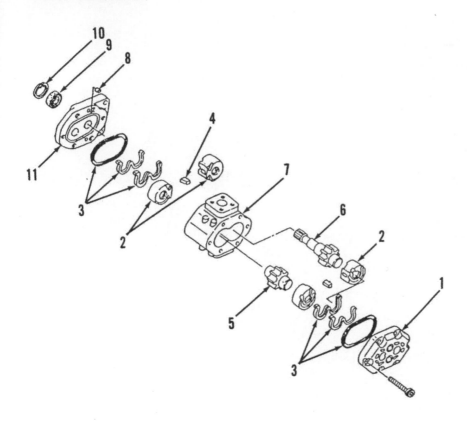

Fig. 118—Exploded view of lift system pump used on 870, 970 and 1070 models.

1. Cover
2. Bushing blocks
3. Packing
4. Key
5. Driven gear
6. Drive gear & shaft
7. Pump body
8. Dowel
9. Oil seal
10. Snap ring
11. Pump flange

Tighten screws attaching inlet and outlet flanges to 6 N·m (53 in.-lbs.) torque.

ROCKSHAFT AND HOUSING

670 And 770 Models

187. REMOVE AND REINSTALL. To remove rockshaft housing, first lower the lift arms and remove any mounted implements. Remove seat and seat support. Remove any rear hydraulic lines which would interfere with removal. Remove lift links from both lift arms. Remove knob from position control lever, then unbolt and remove the lever guide from right fender. Unbolt and remove braces between fenders. Disconnect wiring and hydraulic lines which would interfere with removal of the lift cover, then remove lines and move the rear wiring out of the way. Cover all openings to prevent the entrance of dirt. Remove the eight cap screws, then lift rockshaft from tractor.

Reinstall by reversing removal procedure. Tighten retaining screws and nut to 54 N·m (40 ft.-lbs.) torque. Be sure to reinstall all of the clamps for wiring and hydraulic lines. Check fluid level after assembling and operating hydraulic systems.

188. OVERHAUL ROCKSHAFT AND LIFT ARMS. Refer to paragraph 187 and remove rockshaft cover. Disconnect position control link from feedback arm (28—Fig. 119), then remove the arm (28), washer

(20), lift arms (4 and 19) and retainer washers (5 and 18). Slide rockshaft (14) from crank (10) and splined sleeves (7 and 16).

If new bushings (9 and 13) are installed, press bushings into housing until 7 mm (0.283 inch) below outer surface of housing. Align marks on ends of shaft with marks on crank (10) and lift arms (4 and 19). Install sleeves (7 and 16) with tapered ends toward inside. Lubricate new seal rings (6 and 17), retainers (5 and 18) and lift arms (4 and 19). Install washer (20) and arm (28), then tighten retaining screws to 50 N·m (37 ft.-lbs.) torque. Reconnect position control link to feedback arm (28) and rockshaft control valve, then reinstall rockshaft cover assembly.

189. R&R AND OVERHAUL ROCKSHAFT CONTROL VALVE. Refer to paragraph 187 for removal of the rockshaft housing. The control valve is attached to underside of cover with three socket head screws.

NOTE: Do not disturb locknut (5—Fig. 120) or lower adjusting stop screw (4). Setting is adjusted at factory and may not be successfully accomplished as a service procedure.

Bend locking tab away from screw (1), then remove screw and plate assembly (2, 3, 4 and 5). Withdraw spool and spring assembly (6, 7, 8 and 9) from housing (59). Remove plug (10) and load check valve (12 and 13). Remove plug (14), spring (15) and lowering valve (16, 17 and 18). Insert small Allen wrench or stiff wire

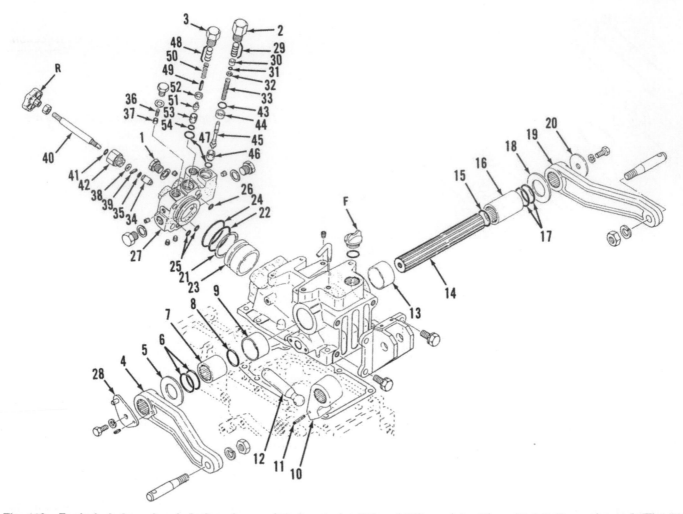

Fig. 119—Exploded view of rockshaft and associated parts for 670 and 770 models. Plugs (1, 2 & 3) are shown in Fig. 111 and Fig. 114.

4. Left lift arm	17. Seal rings (same as 16)	30. Bushing
5. Retainer washer	18. Retainer washer	31. "O" ring
(same as 18)	(same as 5)	32. Washer
6. Seal rings (same as 17)	19. Right lift arm	33. Spring
7. Splined sleeve (left side)	20. Retainer washer	34. Rockshaft drop/
8. "O" ring (same as 15)	21. Back-up ring	lock valve
9. Bushing	22. Seal ring	35. "O" ring
10. Lift crank	23. Piston	36. Spring
11. Roll pin	24. "O" ring	37. Lowering check valve
12. Piston rod	25. "O" rings	38. Washer
13. Bushing	26. "O" ring	39. Pin
14. Rockshaft	27. Cylinder cover	40. Screw
15. "O" ring (same as (8)	28. Feedback arm	41. Packing
16. Splined sleeve (right side)	29. Shims	42. Guide

43. "O" ring
44. Bushing
45. System relief valve
46. Seat
47. Seal ring
48. Shims
49. Roll pin
50. Spring
51. Implement relief
valve poppet
52. Seat retainer
53. Seat
54. "O" ring

into hole for spool (6) to push flow control spool (21) and related parts (19, 20, 22) from bore. Remove plug (23), "O" ring (24), spring (25) and unloading valve (26).

When reassembling, install new "O" rings and coat valves and bores with clean hydraulic fluid. Assemble in reverse of disassembly procedure. Clean threads of screw (1) and spool (6) and use "Loctite" or equivalent to lock threads when assembling. Make sure that cross pin of stop (9) is perpendicular to valve housing mounting surface when screw (1) is tight. Tighten

screw (1) to 10 N·m (88 in.-lbs.) torque, then bend tab of lockplate (2) around flat of screw (1).

Install "O" ring (58) in groove of valve housing (59), then attach valve to rockshaft housing. Tighten the three socket head screws to 26 N·m (230 in.-lbs.) torque.

190. R&R AND OVERHAUL CYLINDER COVER AND PISTON. Lower the lift arms and remove any mounted implements. Remove seat and seat support. Remove the fender brace and any rear

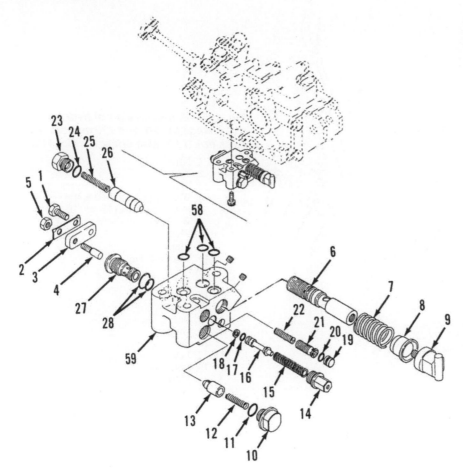

Fig. 120—Exploded view of the rockshaft control valve for 670 and 770 models.

1. Screw
2. Lockplate
3. Connecting plate
4. Adjusting screw
5. Locknut
6. Valve spool
7. Spring
8. Spring guide
9. Stop
10. Plug
11. "O" ring
12. Spring
13. Load check valve
14. Plug
15. Spring
16. Lowering valve
17. "O" ring
18. Back-up ring
19. Plug
20. "O" ring
21. Flow control spool
22. Spring
23. Plug
24. "O" ring
25. Spring
26. Unloading valve
27. Valve seat
28. Seals
58. "O" rings
59. Valve body

hydraulic lines which would interfere with removal. Individual valves can be removed from cylinder cover (27—Fig. 119) for service without removing the cover. Remove the four screws retaining cover, then pull cover from cylinder.

Piston (23—Fig. 119) can be removed from top (front) of cylinder after removing cylinder cover (27). Manually raise the lift arms and push piston into the cylinder against rockshaft piston rod, then push piston from cylinder with a fast downward thrust of the lift arms.

To remove the rockshaft drop/lock valve, unscrew guide (42—Fig. 119) from cylinder cover. Lowering check valve is located at (37) and can be removed after removing plug. The thermo or implement relief valve assembly can be removed after removing plug (3). The system pressure relief valve (45) can be removed after removing plug (2). Be careful not to mix, lose or damage shims (29 and 48) located between springs and plugs.

Install "O" ring (22) and back-up ring (21) in groove of piston, coat rings, piston and cylinder with hydraulic fluid, then insert piston in bore. Make sure that ends of back-up rings are not overlapped and that rings are not damaged while inserting into cylinder.

Coat all parts of valves in clean hydraulic oil before assembling. Be sure to install new "O" rings (47 and 54) if seats (46 and 53) were removed, then tighten retainers (44 and 52).

Before installing cylinder cover, install and lubricate new "O" ring (24), then stick "O" rings (25 and 26) in position with grease. Install cylinder cover, then install and tighten the four retaining screws diagonally and evenly to 95 N•m (70 ft.-lbs.) torque.

191. SELECTIVE CONTROL VALVE. To remove the selective control valve from models so equipped, lower rockshaft arms and lower any equipment attached to hydraulic system. Remove selective control valve lever, rockshaft stop knob, rear pto knob and mid pto knob (if so equipped), then remove cover that surrounds these controls. Remove knob from position control lever (5—Fig. 121), then unbolt and remove the lever guide (20) and the right fender. Disconnect hydraulic lines from selective control valve (10—Fig. 122), then remove screws attaching valve to housing. Cover all openings of hydraulic system immediately after disconnecting lines to prevent the entrance of dirt.

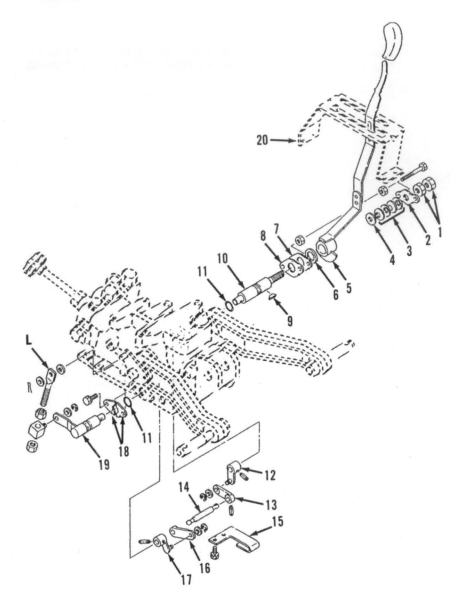

Fig. 121—Exploded view of hydraulic control linkage for 670 and 770 models. Feedback link (L) is also shown in Fig. 116.

1. Nuts
2. Plate
3. Washers
4. Washer (13 × 24 × 2.5 mm)
5. Position control lever
6. Washer
7. Plate
8. Plate
9. Woodruff key
10. Shaft
11. "O" rings
12. Lever
13. Bar
14. Bar
15. Guide
16. Bar
17. Lever
18. Retainers
19. Shaft
20. Guide

Refer to Fig. 123 or Fig. 124 for disassembled view of selective control valve. Valve spools and body are not serviced separately. A new valve assembly should be installed if cleaning does not restore valve to original condition.

870, 970 And 1070 Models

192. REMOVE AND REINSTALL. To remove rockshaft housing, first lower the lift arms and remove any mounted implements. Remove seat, seat support, fuel tank and fenders. Remove any rear hydraulic lines which would interfere with removal. Remove knob from the range shift lever, and remove lever for front wheel drive if so equipped. Remove the cover from left side, around range shift lever. Remove guide from right side, around the draft control lever.

Disconnect spring, linkage and fuel hoses, then remove supports located between fenders. Disconnect wiring and hydraulic lines which would interfere with removal of the lift cover, then remove lines and move the rear wiring out of the way. Cover all openings to prevent the entrance of dirt. Remove 3-point hitch center link and lift links from both lift arms. Remove the twelve cap screws, then lift rockshaft from tractor.

Reinstall by reversing removal procedure. Tighten retaining screws to 88 N·m (65 ft.-lbs.) torque. Be sure to reinstall all of the clamps for wiring and hydraulic lines. Check fluid level after assembling and operating hydraulic systems.

193. OVERHAUL ROCKSHAFT AND LIFT ARMS. Refer to paragraph 192 and remove rockshaft cover. Disconnect position control link from feedback

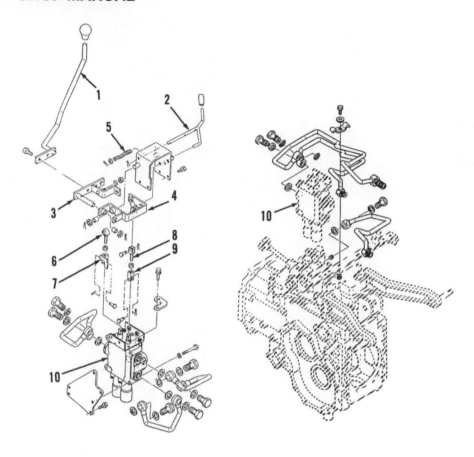

Fig. 122—Exploded view of linkage for selective control valve available for 670 and 770 models. Refer to Fig. 123 or Fig. 124 for exploded view of valve.

1. Control lever
2. Lockout lever
3. Bucket control arm
4. Link arm
5. Spring
6. Rod end
7. Clevis
8. Clevis
9. Clevis
10. Selective control valve

arm (28—Fig. 125), then remove the arm, washer (20), lift arms (4 and 19) and retainer washers (5 and 18). Slide rockshaft (14) from crank (10) and splined sleeves (7 and 16).

If new bushings (9 and 13) are installed, press bushings into housing until 7 mm (0.283 in.) below outer surface of housing. Align marks on ends of shaft with marks on crank (10) and lift arms (4 and 19). Install sleeves (7 and 16) with tapered ends toward inside. Lubricate new spline seals (6 and 17), retainers (5 and 18) and lift arms (4 and 19). Install washer (20) and arm (28), then tighten retaining screws to 88 N·m (65 ft.-lbs.) torque. Reconnect position control link to feedback arm (28) and rockshaft control valve, then reinstall rockshaft cover assembly.

194. R&R AND OVERHAUL ROCKSHAFT CONTROL VALVE. Refer to paragraph 192 for removal of the rockshaft housing. The control valve is attached to underside of cover with three socket head screws.

NOTE: Do not disturb locknut (5—Fig. 126 or Fig. 127) or lower adjusting stop screw (4). Setting is adjusted at factory and may not be successfully accomplished as a service procedure.

Bend locking tab away from screw (1), then remove screw and plate assembly (2, 3, 4 and 5). Withdraw spool and spring assembly (6, 7, 8 and 9) from housing (59). Remove plug (10) and load check valve (12 and 13). Remove holder (14), spring (15) and lowering valve (16, 17 and 18). Insert small Allen wrench or stiff wire into hole for spool (6) to push flow control spool (21) and related parts (19, 20, 22) from bore. Remove plug (23), "O" ring (24), spring (25) and unloading valve (26).

When reassembling, install new "O" rings and coat valves and bores with clean hydraulic fluid. Assemble in reverse of disassembly procedure. Clean threads of screw (1—Fig. 126 or Fig. 127) and spool (6) and use "Loctite" or equivalent to lock threads when assembling. Tighten screw (1) to 10 N·m (88 in.-lbs.) torque, then bend tab of lockplate (2) around flat of screw (1).

Install "O" ring (58) in groove of valve housing (59), then attach valve to rockshaft housing. Tighten the three socket head screws to 26 N·m (230 in.-lbs.) torque.

195. R&R AND OVERHAUL CYLINDER COVER AND PISTON. Lower the lift arms and remove any mounted implements. Remove seat and seat support. Remove knobs from pto control levers and rockshaft stop valve, then remove the front

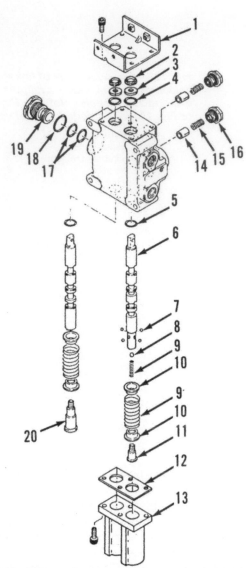

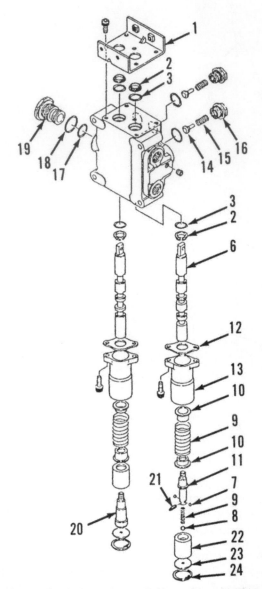

Fig. 123—Exploded view of Danfoss/Dukes selective control valve used on some models. Refer also to Fig. 124.

1. Bracket	11. Guide
2. Seal	12. Gasket
3. Spacer washer	13. Cover
4. Seal	14. Load check valve
5. "O" ring	15. Spring
6. Spool	16. Plug
7. Detent balls	17. "O" rings
8. Ball	18. "O" ring
9. Springs	19. Plug
10. Retainers	20. Guide

Fig. 124—Exploded view of Kanzaki selective control valve used on some models. Refer also to Fig. 123.

1. Bracket	14. Load check valve
2. Seal	15. Spring
3. Spacer washer	16. Plug
6. Spool	17. "O" ring
7. Detent balls	18. "O" ring
8. Ball	19. Plug
9. Springs	20. Guide
10. Retainers	21. Roll pin
11. Retainer	22. Retainer
12. Gasket	23. Cover
13. Cover	24. Retaining ring

panel. Disconnect hydraulic line from cylinder cover and cover all openings to prevent the entrance of dirt. Remove the six screws retaining cover, then pull cover from cylinder.

Piston (23—Fig. 125) can be removed from top (front) of cylinder after removing cylinder cover (27). Manually raise the lift arms and push piston into the cylinder against rockshaft piston rod, then push pis-

ton from cylinder with a fast downward thrust of the lift arms.

To remove the rockshaft drop/lock valve, unscrew guide (42—Fig. 125) from cylinder cover. Lowering check valve is located at (37) and can be removed after removing plug (45).

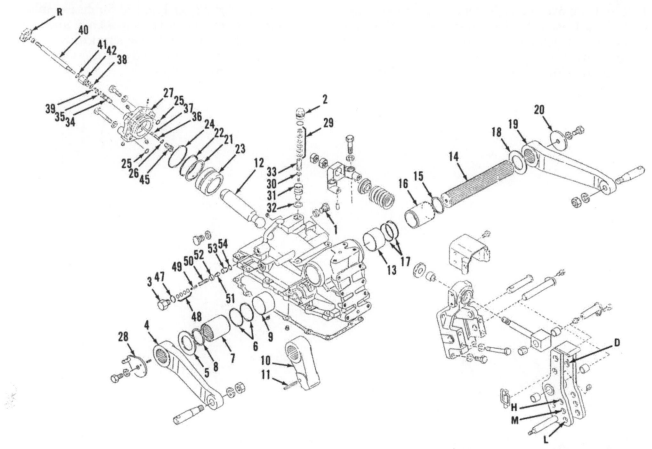

Fig. 125—Exploded view of rockshaft and associated parts typical of 870, 970 and 1070 models. Plugs (1, 2 & 3) are shown in Fig. 112 and Fig. 115. Draft sensing top link shown is only available on 1070 models. To block out draft sensing, pin should be installed in hole (D). Attach upper link in hole (H) for heavy loads, hole (M) for medium loads or hole (L) for light loads.

4. Left lift arm	17. "O" rings (same as 16)	30. System relief valve	41. Packing
5. Retainer washer (same as 18)	18. Retainer washer (same as 5)	31. Seat	42. Guide
6. "O" rings (same as 17)	19. Right lift arm	32. Seal ring	45. Plug
7. Splined sleeve (left side)	20. Retainer washer	33. Spring	47. Seal ring
8. Spline seal (same as 15)	21. Back-up ring	34. Rockshaft drop/	48. Shims
9. Bushing	22. Seal ring	lock valve	49. Roll pin
10. Lift crank	23. Piston	35. "O" ring	50. Spring
11. Pin	24. "O" ring	36. Spring	51. Implement relief
12. Piston rod	25. "O" rings	37. Lowering check valve	valve poppet
13. Bushing	26. Packing ring	38. Washer	52. Seat retainer
14. Rockshaft	27. Cylinder cover	39. Pin	53. Seat
15. Spline seal (same as 8)	28. Feedback arm	40. Screw	54. "O" ring
16. Splined sleeve (right side)	29. Shims		

Install "O" ring (22) and back-up ring (21) in groove of piston, coat rings, piston and cylinder with hydraulic fluid, then insert piston in bore. Make sure that ends of back-up rings are not overlapped and that rings are not damaged while inserting into cylinder. Coat all parts of valves in clean hydraulic oil before assembling. Be sure to install new "O" rings (26, 35 and 41) if removed.

Before installing cylinder cover, install and lubricate new "O" ring (24), then stick "O" rings (25 and 26) in position with grease. Install cylinder cover,

then install and tighten the six retaining screws diagonally and evenly to 95 N•m (70 ft.-lbs.) torque.

196. SELECTIVE CONTROL VALVE. To remove the selective control valve (10—Fig. 128) from models so equipped, lower rockshaft arms and lower any equipment attached to hydraulic system. Remove rockshaft stop knob, rear pto knob and mid pto knob (if so equipped), then remove cover that surrounds these controls. Remove selective control valve lever.

Unbolt and remove the right fender. Disconnect hydraulic lines from selective control valve, then remove screws attaching valve to housing. Cover all openings of hydraulic system immediately after disconnecting lines to prevent the entrance of dirt.

Refer to Fig. 123 or Fig. 124 for disassembled view of selective control valve. Valve body and spools are not available separately for service. New valve assembly should be installed if cleaning does not restore valve to original condition.

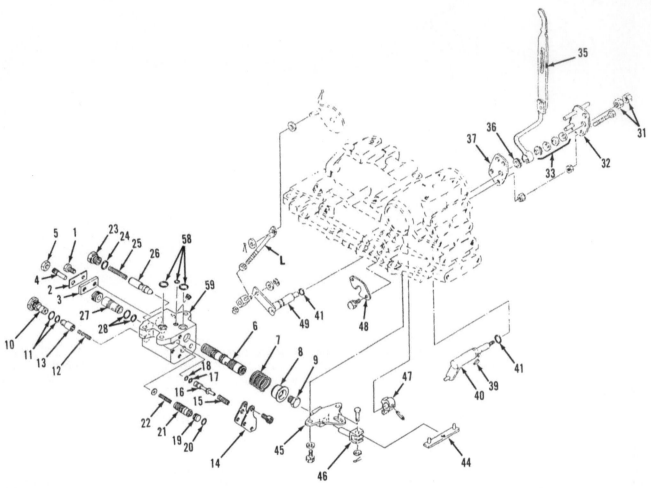

Fig. 126—Exploded view of rockshaft control valve and related parts typical of 870 and 970 models.

1. Screw	13. Load check valve	25. Spring
2. Lockplate	14. Holder	26. Unloading valve
3. Connecting plate	15. Spring	27. Valve seat
4. Adjusting screw	16. Lowering valve	28. "O" rings
5. Locknut	17. "O" ring	29. Shims
6. Valve spool	18. Back-up ring	30. Plug
7. Spring	19. Plug	31. Nuts
8. Spring guide	20. "O" ring	32. Plate
9. Retainer	21. Flow control spool	33. Washers
10. Seat	22. Spring	35. Position control lever
11. "O rings	23. Plug	36. Washer
12. Spring	24. "O" ring	37. Plate

39. Woodruff key
40. Lever & shaft
41. "O" rings
44. Bar
45. Plate
46. Rod
47. Lever
48. Retainer
49. Shaft
58. "O" rings
59. Valve body

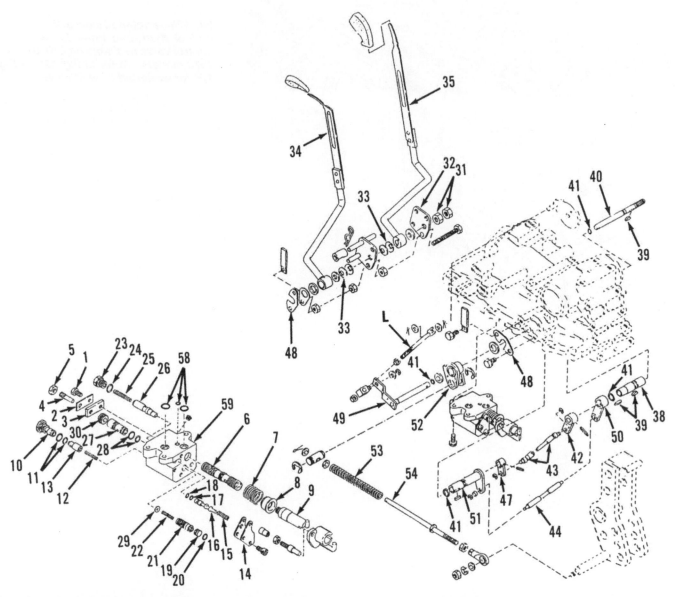

Fig. 127—Exploded view of rockshaft control valve and related parts typical of 1070 model with draft control. Refer to Fig. 126 for legend except the following.

34. Draft control lever
38. Draft control shaft
42. Arm

43. Link
50. Arm

51. Lever & shaft
52. Draft sensing lever

53. Spring
54. Link

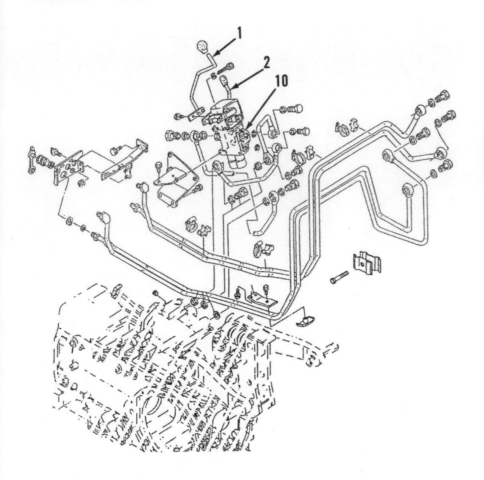

Fig. 128—Exploded view of linkage and typical hydraulic lines for selective control valve available on 870, 970 and 1070 models. Refer to Fig. 123 or Fig. 124 for exploded view of valve.

1. Control lever
2. Lockout lever
10. Selective control valve

NOTES

NOTES

NOTES

NOTES